3 8015 02355 660 2
KU-558-239

Mathematics

Foundation

sion
de

24 MAR 2020
Nov 2nd 2020
DEC 2020
8 – OCT 2019
19 NOV 2019
7 – DEC 2019
28 DEC 2019
18 JAN 2020
8 FEB 2020
24 FEB 2020

Tony Fisher
June Haighton
Andrew Manning
Anthony Staneff
Margaret Thornton

Series Editor
Paul Metcalf

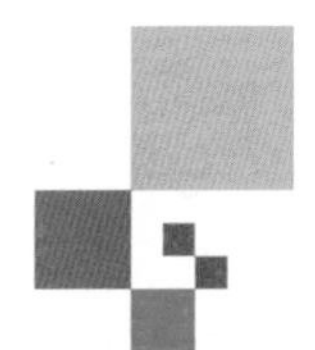

Nelson Thornes

Text © Tony Fisher, June Haighton, Andrew Manning, Anthony Staneff and Margaret Thornton 2010
Original illustrations © Nelson Thornes Ltd 2010

The right of Tony Fisher, June Haighton, Andrew Manning, Anthony Staneff and Margaret Thornton to be identified as the authors of this work has been asserted by them in accordance with the Copyright, Designs and Patents Act 1988.

All rights reserved. No part of this publication may be reproduced or transmitted in any form or by any means, electronic or mechanical, including photocopy, recording or any information storage and retrieval system, without permission in writing from the publisher or under licence from the Copyright Licensing Agency Limited, of Saffron House, 6–10 Kirby Street, London, EC1N 8TS.

Any person who commits any unauthorised act in relation to this publication may be liable to criminal prosecution and civil claims for damages.

Published in 2010 by:
Nelson Thornes Ltd
Delta Place
27 Bath Road
CHELTENHAM
GL53 7TH
United Kingdom

10 11 12 13 14 / 10 9 8 7 6 5 4 3 2

A catalogue record for this book is available from the British Library

ISBN 978 1 4085 0619 6

Cover photographs by iStockphoto (top), PureStock/Photolibrary (middle), Photolibrary (bottom)
Illustrations by Tech-Set Limited
Page make-up by Tech-Set Limited, Gateshead

Printed in China by 1010 Printing International Ltd

Photograph acknowledgements
Fotolia: p10, p11, p30
iStockphoto: p50

CROYDON LIBRARIES

SHR

3 8015 02355 660 2	
Askews	19-Jan-2011
510	£5.99

Contents

How to use this Revision Guide

This book has been written by teachers and examiners to prepare you for your AQA GCSE Mathematics exams. It covers all of the main points that you need to know, and it includes references to the Student Books if you want greater detail. If you are taking the modular course you will be assessed in each of three units. The signpost icon Unit 1 is used throughout this guide to show you what unit the content will be assessed in. If you are taking the linear course, you will be assessed on any of the content in each of two exams, one of which will be non-calculator. The following icon is used throughout this guide to show that you should answer a question without using a calculator :

There are some free, downloadable resources available to accompany this guide. See the back cover for details.

In the exams, you will be tested on the following Assessment Objectives (AOs):

AO1 recall and use your knowledge of the prescribed content

AO2 select and apply mathematical methods in a range of contexts

AO3 interpret and analyse problems and generate strategies to solve them.

You will also be assessed on your Quality of Written Communication (QWC).

This book is split into five sections:

- Number – all the basic number work you need to know. Remember that for some of the work you will not be able to use your calculator.
- Statistics – all of this material is assessed in Unit 1 if you are taking the modular course.
- Algebra – this is covered in Unit 2 and Unit 3 in the modular course. As with the number work, some of this work must be done without using a calculator.
- Geometry and measure – all of this material is assessed in Unit 3 if you are taking the modular course.
- Essential skills – this section looks at the skills that you will need in order to cover QWC, AO2 and AO3 (see above).

Each chapter has the following features:

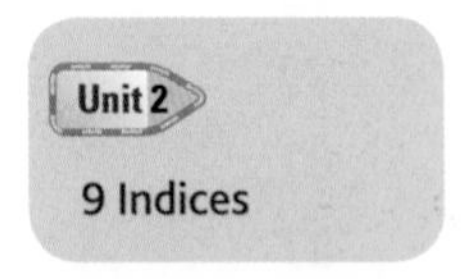

This feature shows you where in the Student Books the material is covered if you need to look at it in more detail.

These are the basic points that you need to understand for the topic.

Key terms

Make sure that you know the meaning of the words in this list by writing down your own definitions and checking them with the glossaries in the Student Books. You are expected to use specialist vocabulary as part of your Quality of Written Communication.

Example

This is an example and worked solution to help show you how to use what you have just revised.

Practise...

As in the Student Books, there are questions that allow you to practise what you have just revised.

The bars that run alongside questions in the exercises show you what grade the question is aimed at. This will give you an idea of what grade you are working at. Do not forget: even if you are aiming at a Grade C, you will still need to do well on the Grades G–D questions.

These questions are Functional Maths type questions, which show how maths can be used in real life.

These questions are problem-solving questions, which will require you to think carefully about how best to answer.

These questions should be attempted with a calculator.

These questions should be attempted without using a calculator.

AQA Examination-style questions

At the end of each section there are some questions in the style of those that you will meet in your exams.

Hint

These are tips for you to remember while learning the maths or answering questions.

AQA Examiner's tip

These are tips from the people who will mark your exams, giving you advice on things to remember and watch out for.

Bump up your grade

There are tips from the people who will mark your exams, giving you help on how to boost your grade. These are especially aimed at getting a Grade C.

Revision tips

WHAT

- Ask your teacher which specification you are following and download a copy from the web. Use this Revision Guide and your exam specification to see what topics you need to cover.
- Go through the topics, past papers and tests and make a list of the areas that you find difficult. Concentrate your revision on these areas.
- Check which formulae will be given on the exam paper. Your teacher can tell you this. Practise using these and make sure that you learn any formulae that are not given.

HOW

- Create an effective revision timetable. A blank timetable spreadsheet is available in the downloadable resources.
  - Find out the date of your exam.
  - Divide your subjects into topics.
  - Mix and match harder and easier topics to break it up a bit.
  - Schedule 10 minutes of top-up time at the start of each session to look back over the work you covered last time.
  - Tick off topics as you complete them. Some checklists are provided in the downloadable resources.
  - Schedule some time off and rewards.
  - Allow time in the last week to do examination-style questions and go over everything one last time.
  - Stick to your schedule, but if you feel comfortable with some topics and are struggling with others, shift it around to allow extra time on the harder topics.
- Use cards to summarise the key points listed in this book. Condense them to one side of paper and take it everywhere with you, reading it at every opportunity.
- Create mind maps or spider diagrams for different topics using plenty of colour, and stick them on your wall. A blank spider diagram is provided in the downloadable resources.
- Do not try to memorise maths; try to understand the processes.
- Use the end-of-chapter and examination-style questions to practise, practise, practise.

Exam tips

- Take the correct equipment. Two pens (only write in blue or black), two sharp pencils, a pencil sharpener, a ruler, an eraser, your calculator with spare batteries, protractor, compasses and a watch to time your answers.
- Things you need to know about *your* calculator before you sit your exam:
  - check that your calculator follows BIDMAS rules (try typing $20 - 9 \times 2 =$; if your calculator gives the answer 2 it is following the rules, if it gives the answer 22 it is not and it is worth getting a newer calculator)
  - how to use brackets
  - how to use the memory
  - how to use the Ans button to insert the previous answer (not all calculators have this function)
  - how to find a square root ($\sqrt{\ }$)
  - how to find a power of a number (3^2, 5^4, ...)
  - how to do fractions
  - how to enter and read figures in standard form.
- Lower-graded, easier questions are at the beginning of the paper, so start with those to ease your way in.
- Pace yourself so that you do not run out of time, and try to allow 10 minutes at the end for checking your answers.
- Try to work neatly and set out your answer clearly in the given space.
- Look at the marks available for a question. A question worth 1 mark does not need lots of explanation or working, but a 4-mark question is likely to need more than one step to get the answer.
- Show all of your working for questions worth more than 1 mark. If you simply write down the answer you are taking a chance that it is completely correct. The examiner will not be able to award you part marks when your method is not shown.
- Use appropriate calculator methods on the calculator paper. For example, build up and long multiplication are methods that are best suited to the non-calculator paper.
- Read the question carefully and interpret the 'exam speak':

Exam speak	What it means
Write down	Working is not needed.
Show; You **must** show your working; Explain/Justify/Support your answer; Give a reason for your answer	Show your method; you will not get any marks otherwise. You can use words, numbers or algebra.
Estimate	Round the numbers, then do the calculation. Do not find the exact answer.
Work out/Calculate/Find	Do a calculation; do not measure.
Measure	Use a ruler or protractor; do not calculate.
Not drawn accurately	The diagram is not accurate; do not measure it.
Use the graph to estimate/solve	Your answer must come from the graph.
Give your answer to a suitable degree of accuracy	Use the same or less accuracy than the numbers given in the question.
State the units of your answer	There will be a special mark for this.

- Check that your answer is sensible. For example, if it is a probability is it between 0 and 1?
- If the question asks you to give your answer to a degree of accuracy then make sure you do.
- Do not round until the end of the question (unless it is an estimating question) or you may lose accuracy.
- Study the examination-style answers to see what the examiner is looking for.

GOOD LUCK and DON'T PANIC!

Number

1 Types of numbers

Unit 1
1 Fractions

Unit 2
1 Types of numbers

Unit 3
1 Fractions and decimals

Key terms

Write down definitions for the following words. Check your answers in the glossary of your Student Book.

BIDMAS
common factor
difference
directed number
factor
highest common factor (HCF)
index
integer
inverse operation
least common multiple (LCM)
multiple
negative number
place value
positive number
prime number
product
quotient
sum

Revise... Key points

In Units 1 and 3 you will be allowed to use a calculator.
In Unit 2 you will have to carry out calculations without a calculator.

Place value (Unit 2)

The value of each digit in a number depends on its position.
When you **multiply a number by 10**, all the digits move **1 place to the left**.
When you **divide a number by 10**, all the digits move **1 place to the right**.
The **place value** table shows twenty-seven thousand and forty, 27 040.

	Thousands						
	H	T	U	H	T	U	
		2	7	0	4	0	← 'twenty-seven thousand and forty'
'27 040 × 10 =' →	2	7	0	4	0	0	← 'two hundred and seventy thousand, four hundred'
'27 040 ÷ 10 =' →			2	7	0	4	← 'two thousand, seven hundred and four'

The value of the digit 2 is 20 000. The value of the digit 7 is 7000. The value of the digit 4 is 40.

Working with whole numbers (All Units 1 2 3)

You can use a calculator in Unit 1 and Unit 3. In Unit 2 you may need to add or subtract numbers of any size **without** a calculator. You may also be asked to multiply a three-digit **integer** by a two-digit integer or divide a three-digit integer by a two-digit integer. Use standard methods unless you spot quicker ways.

Use **inverse operations** to check your answers.
Addition and subtraction are inverses. Multiplication and division are inverses.

Link
Unit 2 Chapter 1 (Types of numbers) includes the standard methods for adding, subtracting, multiplying and dividing whole numbers.

BIDMAS (All Units 1 2 3)

BIDMAS can help you to remember the correct order to work things out.

B	I	D	M	A	S
Brackets	**Indices**	**Divide**	**Multiply**	**Add**	**Subtract**

Divide and Multiply: These go together. When they are both in a calculation, work from left to right.

Add and Subtract: These go together. When they are both in a calculation, work from left to right.

For example, $8 - 3 \times 2 = 8 - 6 = 2$ ← Multiply before subtracting

Check whether your calculator uses BIDMAS. If not, you will need to remember the correct order in the exams for Unit 1 and Unit 3 as well as Unit 2.

Adding and subtracting directed numbers

All Units 1 2 3

Use a number line to put **directed numbers** in order. It can also help you to add or subtract directed numbers when you are not allowed to use a calculator (in Unit 2).

To **add directed numbers**, find the first number on the number line.

Go right to add a **positive number** or left to add a **negative number**.

The green arrows show $-45 + 22 = -23$ and $20 + -28 = -8$

Adding finds the **sum** of numbers.

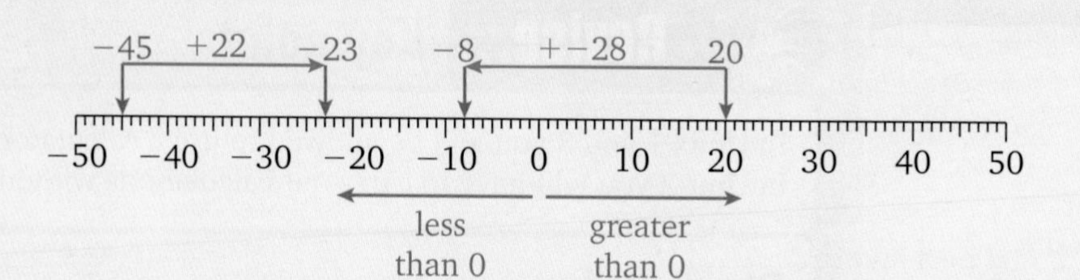

To **subtract directed numbers**, find the first number on the number line.

Go left to subtract a positive number or right to subtract a negative number.

The black arrows show $-17 - 23 = -40$ and $-5 - -25 = 20$

Subtracting finds the **difference** between numbers.

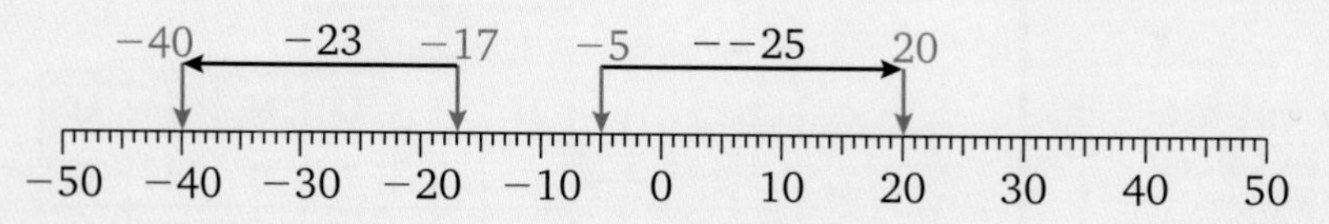

Adding a positive number + +	does the same as	+
Adding a negative number + −	does the same as	−
Subtracting a positive number − +	does the same as	−
Subtracting a negative number − −	does the same as	+

Hint

Make sure that you know how to enter negative numbers on your calculator for Unit 1 and Unit 3.

Multiplying and dividing directed numbers

All Units 1 2 3

To multiply or divide directed numbers without a calculator, use the following rules:

When signs are the same	+ × + + ÷ +	or or	− × − − ÷ −	the answer is +
When signs are different	+ × − + ÷ −	or or	− × + − ÷ +	the answer is −

For example, $5 \times -4 = -20$ and $-5 \times -4 = 20$

$-12 \div 4 = -3$ and $-12 \div -4 = 3$

Multiplying finds the **product** of numbers.

Dividing finds the **quotient**.

Factors and multiples

Unit 2

To find all the **factors** of a number, look for factor pairs.

To find the **common factors** of two numbers, list all their factors.
Then look for the numbers that appear in both lists.
The **highest common factor (HCF)** is the **biggest** number that appears in both lists.

To find the **least common multiple (LCM)** of two numbers, list the **multiples** of each number.
The least common multiple is the **smallest** number that appears in both lists.

Bump up your grade

For Grade C you should be able to find highest common factors and least common multiples.

Prime factors Unit 2

Use the tree method to write a number as the **product** of its prime factors.

Here is one of the ways of splitting up 120. It does not matter which pair of factors you start with, you will always end up with the same prime factors.

The tree shows that

$120 = 2 \times 2 \times 2 \times 3 \times 5$

In **index** form $120 = \mathbf{2^3 \times 3 \times 5}$

120 → 4, 30; 4 → (2), (2); 30 → 6, (5); 6 → (3), (2)

Keep splitting up the numbers until you reach **prime numbers**.

Alternatively, you could start the tree with the factors 2 and 60. (Remember 2 is a factor of all even numbers.)

Bump up your grade

For a Grade C you must be able to write a number as a product of prime factors.

Example Working with whole numbers 

F E

 The table gives the distances in nautical miles between some ports in the UK.

Dover				
260	Falmouth			
63	322	Felixstowe		
204	464	160	Hull	
120	160	183	324	Southampton

a How much further is it from Felixstowe to Falmouth than from Felixstowe to Southampton?

b A ship sails from Felixstowe to Southampton, from Southampton to Hull, then from Hull back to Felixstowe.
The ship sails 12 nautical miles every hour.
Work out how many hours the whole journey takes. Give your answer to the nearest hour.

Solution

a The distance from Felixstowe to Falmouth is 322 nautical miles.
The distance from Felixstowe to Southampton is 183 nautical miles.
It is 139 nautical miles further to Falmouth.

$$\begin{array}{r} {}^{2}\cancel{3}\;{}^{11}\cancel{2}\;{}^{1}2 \\ -\;1\;8\;3 \\ \hline 1\;3\;9 \end{array}$$

AQA Examiner's tip

When subtracting numbers, remember to line up the digits with the larger number on top.

b The distance from Felixstowe to Southampton is 183 nautical miles.
The distance from Southampton to Hull is 324 nautical miles.
The distance from Hull back to Felixstowe is 160 nautical miles.
The total distance is 667 nautical miles.

$$\begin{array}{r} 1\;8\;3 \\ 3\;2\;4 \\ +\;1\;6\;0 \\ \hline 6\;6\;7 \\ {}_{1}\;\;\;\;\; \end{array}$$

The ship travels 12 nautical miles each hour.
To find the time for the whole journey, divide 667 by 12.

$$12\overline{)6\;6\;{}^{6}7}\quad = 5\;5\;\text{r}7$$

$1 \times 12 = 12$
$2 \times 12 = 24$
$3 \times 12 = 36$
$4 \times 12 = 48$
$5 \times 12 = 60$

The remainder of 7 is more than half of 12. This means the answer is nearer 56 than 55.

The time for the whole journey is 56 hours, to the nearest hour.

Example BIDMAS Unit 2

E

Insert brackets into this statement to make it correct $4 + 2 \times 5 - 7 = -12$

Solution

The result of the calculation is -12.

Look for ways of inserting brackets to give a negative value.

Remember to use BIDMAS and the rules for positive and negative numbers.

Doing $5 - 7$ first would give -2, so try putting brackets around $5 - 7$

$4 + 2 \times (5 - 7) = 4 + 2 \times -2 = 4 - 4 = 0$ so this is not correct.

Note that a bracket around $4 + 2$ here would give $6 \times -2 = -12$

So, put another bracket in the statement to give $(4 + 2) \times (5 - 7) = 6 \times -2 = -12$

Example Factors and multiples  Unit 2

C

a **i** Find all the common factors of 20 and 36.

ii What is the highest common factor of 20 and 36?

b Alice, Laura and Chloe all swim regularly at the same swimming pool.
Alice swims every 5 days, Laura swims every 6 days and Chloe swims every 10 days.
They all swam today. After how many days will they all swim again on the same day?

Solution

a **i** $20 = 1 \times 20$ or 2×10 or 4×5, so the factors of 20 are **1**, **2**, **4**, 5, 10, 20.

$36 = 1 \times 36$ or 2×18 or 3×12 or 4×9 or 6×6,
so the factors of 36 are **1**, **2**, 3, **4**, 6, 9, 12, 18, 36.

The common factors of 20 and 36 are 1, 2 and 4.

ii The highest common factor of 20 and 36 is **4**.

b Alice swims again after 5 days, 10 days, 15 days, 20 days, 25 days, **30 days**, … These are the multiples of 5.

Laura swims again after 6 days, 12 days, 18 days, 24 days, **30 days**, …

Chloe swims again after 10 days, 20 days, **30 days**, …

They all swim again on the same day after 30 days. 30 is the least common multiple of 5, 6 and 10.

Practise... Types of numbers  All Units 1 2 3

Answer all questions without a calculator. Then use a calculator to check the answers where possible.

G

1 The number of spectators at a football match is 23 672.

a Write the number 23 672 in words.

b In the number 23 672, write down the value of:

i the digit 7 **ii** the digit 3.

c Last week there were 21 987 spectators.
Which number is greater, 23 672 or 21 987?

G F

2 **a** Here is a list of numbers: 5070 7050 5700 5007 7500 7005

i From the list, write down the smallest number.

ii From the list, write down the largest number.

iii From the list, write down the number that is nearest to 6300.

b Copy and complete the following:

i $7500 \times 10 = \_$ **ii** $7500 \div 10 = \_$

3 **a** Work these out:

i $79 + 648$ **ii** $900 - 284$ **iii** 37×26 **iv** $642 \div 6$

b Use inverse operations to check your answers.

4 The outside temperature, in °C, at a zoo at 6 am on five winter days was recorded.

Day	Monday	Tuesday	Wednesday	Thursday	Friday
Temperature	−4	−2	−6	−1	0

a Which day was the warmest at 6 am?

b Which day was the coldest at 6 am?

c The temperature rose by 7 °C between 6 am and noon on Tuesday.
What was the temperature at noon on Tuesday?

F E

5 Copy these and insert brackets to make them correct.

a $8 - 4 + 1 = 3$

b $5 + 2 - 4 \times 2 = 6$

c $8 + 6 \div 3 - 1 = 7$

d $9 - 6 \times 4 \div 3 = -5$

E

6 Ian says, 'When you subtract a negative number from a negative number you always get a positive answer.' Give an example to show that Ian is **not** correct.

G E

7 When going to work, I travel on bus number 27 then bus number 43.
Say whether each of the following statements is true or false.
Give a reason for each answer.

a 27 and 43 are both prime numbers.

b The sum of 27 and 43 is an odd number.

c 3 is a factor of 43

d 43 minus 27 is a multiple of 4.

G D

8 Here is a list of numbers: 9 15 19 24 27 45 56 58
From this list, write down

a a multiple of 7

b two factors of 54

c a square number

d a prime number

e a number that is a common multiple of 3 and 4.

D

9 Write down all the common factors of 12 and 18.

C

10 **a** Write down all the factors of **i** 14 **ii** 35

b What is the highest common factor (HCF) of 14 and 35?

c Find the least common multiple (LCM) of 14 and 35.

C

11 a Write 54 as the product of its prime factors. Give your answer in index form.

b Find the highest common factor (HCF) of 54 and 36.

12 p is a prime number. Jo says $3p$ is always odd. Give an example to show Jo is wrong.

13 p is an odd number.

a Is $2p - 1$ an odd number, an even number or could it be either?

b Explain why $p^2 + 1$ is always an even number.

14 Seth wants to spend a week in a city. The table gives the charge per night at two hotels.
Which hotel is the cheapest for the whole week? Give a reason for your answer.

	Price per night (£)	
Nights	**Budget Hotel**	**Classic Hotel**
Mon, Tues, Wed, Thurs	78	82
Fri, Sat, Sun	59	53

15 Here is part of Rowan's bank statement.

Date	Details	Withdrawn	Paid in	Balance
22 Nov	Starting balance			−£45
23 Nov	Cheque number 002377	£36		
23 Nov	Cash paid in		£120	
25 Nov	Cheque number 002378	£75		
27 Nov	Monthly salary		£982	

Copy and complete the Balance column to show how much Rowan has in the account after each transaction.

16 Kate makes writing sets to sell at the school fair.
Each set contains a notepad, pencil and pencil sharpener.
Kate buys the notepads, pencils and pencil sharpeners in packs.
There are 10 pencils in each pack.
There are 6 notepads in each pack.
There are 4 pencil sharpeners in each pack.

Kate needs **exactly** the same number of pencils, notepads and pencil sharpeners.

What is the smallest number of each pack she must buy? You **must** show your working.

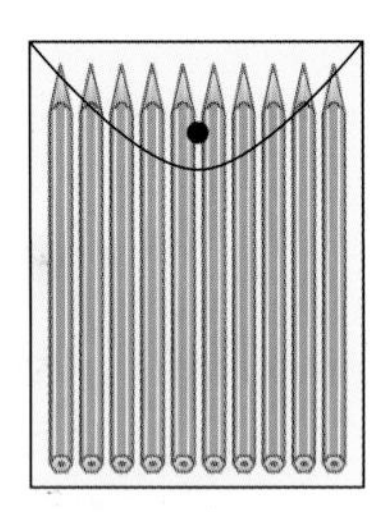
10 pencils

6 notepads

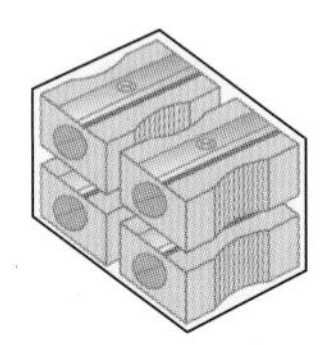
4 pencil sharpeners

Number

2 Fractions, decimals and rounding

Key terms

Write down definitions for the following words. Check your answers in the glossary of your Student Book.

decimal
decimal places
denominator
digit
equivalent fraction
improper fraction
integer
mixed number
numerator
reciprocal
recurring decimal
round
significant figures
terminating decimal

Revise... Key points

Fractions and mixed numbers

All Units 1 2 3

You can think of a simple fraction like $\frac{3}{5}$ in different ways, for example:

- a position on a number line between 0 and 1
- 3 parts out of 5
- 3 divided by 5
- equivalent to the **decimal** 0.6.

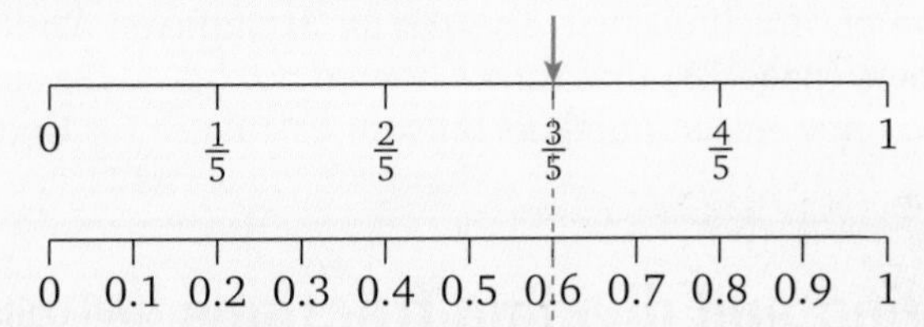

Mixed numbers like $2\frac{3}{5}$ can be written as improper (top-heavy) fractions:

$$2\frac{3}{5} = 2 \times \frac{5}{5} + \frac{3}{5} = \frac{13}{5}$$

The **denominator** of the **improper fraction** is the same as the denominator in the fraction part of the mixed number. To find the **numerator**, multiply the **denominator** by the **integer** part of the mixed number and add the numerator of the fraction. (This changes all of $2\frac{3}{5}$ into fifths.)

$2\frac{3}{5}$ is also equal to the decimal 2.6.

Equivalent fractions

All Units 1 2 3

The value of a fraction does not change when you multiply or divide the numerator and denominator by the same number. Doing this gives an **equivalent fraction**.

To compare the size of fractions like $\frac{1}{3}$ and $\frac{2}{7}$, write them with the same denominator.

Look for the smallest number that is a multiple of each of the denominators. In this case use 21.

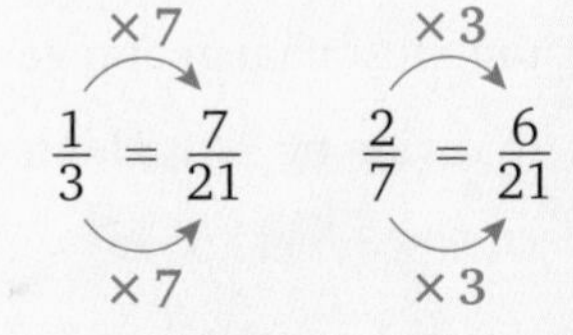

$\frac{7}{21}$ is bigger than $\frac{6}{21}$, so $\frac{1}{3}$ is bigger than $\frac{2}{7}$.

To simplify a fraction, divide the numerator and denominator by the same numbers until you find the simplest equivalent form. Sometimes there is more than one way to do this. Here are two ways to simplify $\frac{24}{30}$:

$$\frac{24}{30} \overset{\div 6}{=} \frac{4}{5} \qquad \frac{24}{30} \overset{\div 2}{=} \frac{12}{15} \overset{\div 3}{=} \frac{4}{5}$$

To simplify $\frac{24}{30}$ on a calculator enter  24 ►30 =

or 24 $a\frac{b}{c}$ 30 =

Make sure that you know how to enter fractions on your calculator. You can use your calculator in Units 1 and 3, but not in Unit 2.

Hint

If the numerator and the denominator are even you can always divide by 2.

Adding and subtracting fractions

All Units 1 2 3

To add or subtract fractions without a calculator (Unit 2), write them with the same denominator.

To add $\frac{2}{5}$ and $\frac{1}{4}$ use a denominator of 20 as it is the lowest common multiple of 5 and 4:

$$\frac{2}{5} + \frac{1}{4} = \frac{8}{20} + \frac{5}{20} = \frac{13}{20}$$

×4 ×5

×4 ×5

In Unit 1 and Unit 3 you can work with fractions on a calculator.
Use the fraction key to enter the fractions and use the normal +, –, × and ÷ keys to do calculations.

To add $\frac{2}{5}$ and $\frac{1}{4}$ press [fraction key] 2 ► 5 ► + [fraction key] 1 ► 4 =

or 2 [$a\frac{b}{c}$] 5 + 1 [$a\frac{b}{c}$] 4 =

Your calculator should give the final answer $\frac{13}{20}$.

To add or subtract mixed numbers without a calculator (Unit 2), change them to improper fractions first. There is an Example on working with mixed numbers (page 18) that shows how to do this. Make sure you also know how to enter mixed numbers on your calculator for Units 1 and 3.

Bump up your grade

In order to get a Grade C you should be able to add and subtract mixed numbers.

Multiplying and dividing fractions

All Units 1 2 3

To multiply fractions without a calculator (in Unit 2), multiply the numerators and multiply the denominators.

For example, $\frac{3}{5} \times \frac{4}{9} = \frac{12}{45} = \frac{4}{15}$ (÷3 top and bottom)

Simplifying first makes the working easier: $\frac{\not{3}^{1}}{5} \times \frac{4}{\not{9}_{3}} = \frac{4}{15}$

AQA Examiner's tip

Make sure that you do not mix up the methods for working with fractions.

Change integers and mixed numbers to improper fractions before multiplying.

In Unit 1 and Unit 3 you can use a calculator. To work out $6 \times 1\frac{3}{4}$ press:

6 × Shift [fraction key] 1 ► 3 ► 4 ► = or 6 × 1 [$a\frac{b}{c}$] 3 [$a\frac{b}{c}$] 4 = The answer is $10\frac{1}{2}$.

To divide by a fraction without a calculator (Unit 2), multiply by its **reciprocal** (the upside-down fraction).

Remember that the reciprocal of any fraction $\frac{a}{b}$ is $\frac{b}{a}$ because $\frac{a}{b} \times \frac{b}{a} = 1$

So $\frac{3}{7} \div \frac{2}{5} = \frac{3}{7} \times \frac{5}{2} = \frac{15}{14} = 1\frac{1}{14}$

(In Units 1 and 3 you can do $\frac{3}{7} \div \frac{2}{5}$ on your calculator.)

AQA Examiner's tip

Remember that only the second fraction is turned upside down.

You must change integers and mixed numbers to improper fractions first. $2\frac{3}{5} = \frac{13}{5}$, so its reciprocal is $\frac{5}{13}$.

To divide by $2\frac{3}{5}$ without a calculator, multiply by $\frac{5}{13}$

The reciprocal of any number x (integer, decimal or fraction) is $\frac{1}{x}$ because $x \times \frac{1}{x} = 1$

You can use the $\frac{1}{x}$ or x^{-1} key on your calculator to find reciprocals in Units 1 and 3.

Finding a fraction of a quantity

All Units 1 2 3

- Divide the quantity by the denominator of the fraction.
- Then multiply by the numerator of the fraction.

So $\frac{3}{5}$ of 20 = 20 ÷ 5 × 3 = 12

AQA Examiner's tip

Make sure you can do this both with a calculator (Units 1 and 3) and without a calculator (Unit 2).

Writing one quantity as a fraction of another

All Units 1 2 3

- Change both quantities into the same units first (if necessary).
- Write the first quantity as the numerator and the second as the denominator of a fraction.
- Simplify the fraction.

Decimal place value

All Units 1 2 3

A decimal point separates the integer part of a number from the fraction part.
You can use a place value table to compare the size of decimal numbers.

The first number in the table below is 47.139. The value of the **digit** 4 is 40. The value of the digit 1 is 0.1 or $\frac{1}{10}$.
The value of the digit 3 is 0.03 or $\frac{3}{100}$. The value of the digit 9 is 0.009 or $\frac{9}{1000}$.
The most significant figure in 47.139 is 4 since this is the digit with the highest value.

Thousands	Hundreds	Tens	Units	.	Tenths	Hundredths	Thousandths
		4	7	.	1	3	9
		4	7	.	6		
		4	7	.	1	8	2

The numbers in the table are 47.139, 47.6 and 47.182.

To put them in order of size, compare the most **significant figures** first. In this case all of the numbers start with 47.
Compare the tenths next. The number 47.6 has 6 tenths, but the others have just 1 tenth. So 47.6 is the largest number.

Now compare the hundredths. 47.139 has 3 hundredths and 47.182 has 8 hundredths, so 47.182 is bigger than 47.139.

Putting the numbers in order of size, starting with the largest, gives 47.6, 47.182, 47.139.

Rounding and estimating

All Units 1 2 3

You might be asked to **round** a number to the nearest ten, hundred or thousand. For example, a population of 24 683 is 24 680 to the nearest ten, 24 700 to the nearest hundred or 25 000 to the nearest thousand.

You may also need to round a decimal to the nearest integer or to 1 **decimal place**, 2 decimal places or 3 decimal places.

A number line can help. This one shows that 2.865 is nearer to 3 than 2. So 2.865 rounded to the nearest integer is 3.

2.865 is nearer to 2.9 than 2.8. So 2.865, rounded to 1 decimal place is 2.9.

2.865 is exactly halfway between 2.86 and 2.87. When it is rounded to 2 decimal places, it is rounded up to 2.87.

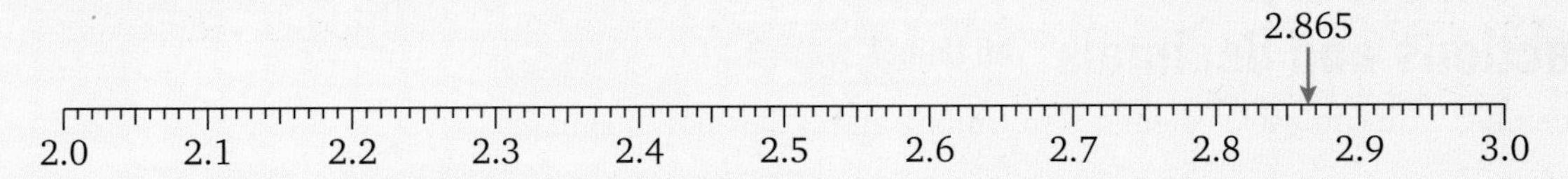

Sometimes you may be asked to round to 1 significant figure.
752 rounded to 1 significant figure is 800. (The zeros at the end of 800 are not significant.)
16.31 rounded to 1 significant figure is 20.
4.92 rounded to 1 significant figure is 5.

Rounding to 1 significant figure can be used to give a rough estimate of the answer to a calculation.

The answer to $\frac{752}{16.3 \times 4.92}$ is approximately equal to $\frac{800}{20 \times 5} = \frac{800}{100} = 8$

AQA Examiner's tip

Be careful not to confuse decimal places and significant figures. For example, 53.76 is 53.8 correct to 1 decimal place, but it is 50 correct to 1 significant figure.

Adding and subtracting decimals

All Units 1 2 3

You will be allowed to use a calculator in Units 1 and 3, but not in Unit 2.

To add (or **subtract)** decimals without a calculator, remember to **line up the decimal points.**

For example 4.8 − 1.56 = 3.24

$$\begin{array}{r} 4\,.\,{}^{7}\not{8}\,{}^{1}0 \\ -\;1\,.\,5\;6 \\ \hline 3\,.\,2\;4 \\ \hline \end{array}$$

Link

Unit 2 Learn 4.3 (Adding and subtracting decimals) shows how to add and subtract decimals without a calculator.

Multiplying decimals

All Units 1 2 3

Remember, you can use a calculator in Units 1 and 3, but not in Unit 2.

To multiply decimal numbers without a calculator:

- Remove the decimal points.
- Multiply the numbers as usual.
- Use an estimate to find out where to put the decimal point. Or count the **total** number of decimal places in the original numbers as the answer should have the same number of decimal places.

For example, 1.2 × 0.03 = 0.036
because 12 × 3 = 36 and $1.\underline{2} \times 0.\underline{0}\underline{3}$ has 3 decimal places.

Link

Unit 2 Learn 4.4 (Multiplying decimals) shows more on how to multiply decimals without a calculator.

Dividing decimals

All Units 1 2 3

Remember, you can use a calculator in Units 1 and 3, but not in Unit 2.

To divide decimal numbers without a calculator:

- Write the division as a fraction.
- Multiply the numerator and denominator so that the denominator is a whole number.
  If the denominator has 1 decimal place, you will need to multiply by 10.
  If the denominator has 2 decimal places, you will need to multiply by 100.
- Divide the numerator by the denominator to find the answer.

For example $1.2 \div 0.03 = \frac{1.2}{0.03} = \frac{120}{3} = 40$

Bump up your grade

For a Grade C you need to be able to divide by decimals.

Link

Unit 2 Learn 4.5 (Dividing decimals) shows more on how to divide decimals without a calculator.

Fractions and decimals

Remember, you can use a calculator in Units 1 and 3, but not in Unit 2.

To change a fraction to a decimal, divide the numerator by the denominator.

Sometimes you get a **terminating decimal** and sometimes you get a **recurring decimal**.

$\frac{1}{2} = 0.5$ is a terminating decimal

$\frac{1}{3} = 1 \div 3 = 0.33\ldots$ is the recurring decimal $0.\dot{3}$

$$\begin{array}{r} 0.\,3\;3\ldots \\ 3\overline{)1.{}^{1}0{}^{1}0\ldots} \end{array}$$

Changing fractions to decimals gives an alternative method for putting fractions in order.

To change a decimal to a fraction, use the place value of the least significant digit as the denominator.

$0.59 = \frac{59}{100}$ because the least significant digit in 0.59 is 9 (**hundredths**)

Example Calculating with decimals

F E

The cost of some fruit at a greengrocer's is shown.

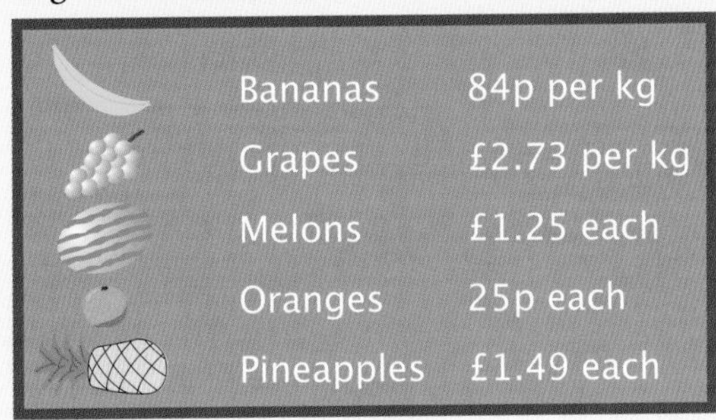

Bananas	84p per kg
Grapes	£2.73 per kg
Melons	£1.25 each
Oranges	25p each
Pineapples	£1.49 each

a Lynn buys 1.5 kg of bananas and 0.4 kg of grapes.
How much change does she get from a £5 note?

b Tom has £10 to spend on pineapples.
What is the maximum number of pineapples Tom can buy?

c Gail buys more than one melon and some oranges.
She spends £4.50.
What could she have bought? Give **two** possible answers.

Solution

a Cost of bananas $= 1.5 \times 84\text{p} = 126\text{p} = £1.26$
Cost of grapes $= 0.4 \times £2.73 = £1.092$
$= £1.09$ to the nearest penny
Total cost $= £1.26 + £1.09 = £2.35$
Change $= £5 - £2.35 = £2.65$

If your calculator gives answers as fractions or mixed numbers, press the S⇔D key to convert them to decimals.

b To find the number of pineapples that Tom can buy, work out $£10 \div £1.49 = 6.71\ldots$
Tom can buy 6 pineapples.

c 2 melons cost $2 \times £1.25 = £2.50$ Gail could then spend $£4.50 - £2.50 = £2$ on oranges.
The number of oranges $= £2 \div £0.25$ (or $200\text{p} \div 25\text{p}$) $= 8$

3 melons cost $3 \times £1.25 = £3.75$ Gail could then spend $£4.50 - £3.75 = £0.75$ on oranges.
The number of oranges $= £0.75 \div £0.25$ (or $75\text{p} \div 25\text{p}$) $= 3$

4 melons cost $4 \times £1.25 = £5$, but this is more than £4.50.

Gail could have bought 2 melons and 8 oranges or 3 melons and 3 oranges.

Example Finding a fraction of a quantity

Unit 2

D

Neil sets out on a journey of 350 miles. His car can travel 9 miles on each litre of petrol.
The petrol tank holds 52 litres when it is full.
The diagram shows how much is in the tank when Neil sets off.

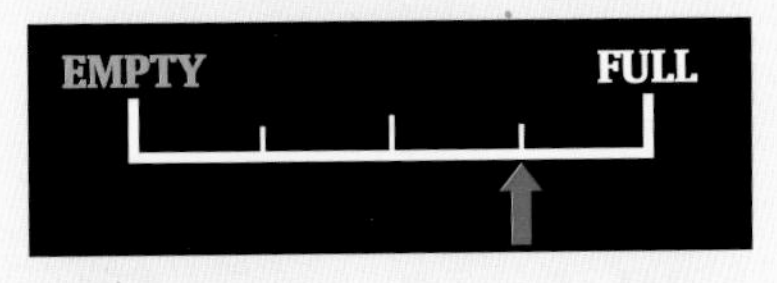

a How far can Neil's car travel on this petrol?

b Should Neil buy more petrol on the way? Explain your answer.

Solution

a The petrol tank is $\frac{3}{4}$ full.
The amount of petrol in the tank is $\frac{3}{4}$ of 52 litres $= 52 \div 4 \times 3 = 13 \times 3 = 39$ litres
The distance Neil can travel using this petrol is $39 \times 9 = 351$ miles

b Yes – Neil should buy more petrol. There is not much spare petrol and if there is heavy traffic or a diversion, Neil's car may need more. He will also need some to start his next journey.

Example Writing one quantity as a fraction of another

D

Write 45 minutes as a fraction of $1\frac{1}{4}$ hours.

Solution

$1\frac{1}{4}$ hours = 60 minutes + 15 minutes = 75 minutes

The fraction is

$$\frac{45}{75} \overset{\div 5}{=} \frac{9}{15} \overset{\div 3}{=} \frac{3}{5}$$

45 minutes is $\frac{3}{5}$ of $1\frac{1}{4}$ hours.

Example Working with mixed numbers

C

A dressmaker makes pyjamas for a children's clothes shop.
Each set of pyjamas consists of a top and trousers.
The dressmaker uses $1\frac{5}{6}$ metres of fabric for each top and $1\frac{1}{2}$ metres for each pair of trousers.
How many sets can she make from a 25-metre roll of fabric?

Solution

The total amount of fabric for one set $= 1\frac{5}{6} + 1\frac{1}{2}$

$= \frac{11}{6} + \frac{3}{2}$

$= \frac{11}{6} + \frac{9}{6}$

$= \frac{20}{6} = \frac{10}{3}$

Number of sets she makes from 25 metres $= 25 \div \frac{10}{3}$

$= \frac{\overset{5}{\cancel{25}}}{1} \times \frac{3}{\underset{2}{\cancel{10}}}$

$= \frac{15}{2} = 7\frac{1}{2}$

The dressmaker can make 7 sets of pyjamas.

Make sure that you also know how to work with mixed numbers on your calculator (for Unit 1 and Unit 3).

To do this example on a calculator press:

Shift [▭/▭] 1 ► 5 ► 6 ► + Shift [▭/▭] 1 ► 1 ► 2 =

then 25 ÷ Ans =

or 1 [$a\frac{b}{c}$] 5 [$a\frac{b}{c}$] 6 + 1 [$a\frac{b}{c}$] 1 [$a\frac{b}{c}$] 2 =

then 25 ÷ Ans =

Your calculator should give $7\frac{1}{2}$

Example Working with fractions and decimals

C

A water tank is $\frac{4}{5}$ full. After 7.5 litres of water are used from the tank it is $\frac{3}{4}$ full.
How much does the tank hold when it is full?

Solution

The fraction that was used $= \frac{4}{5} - \frac{3}{4} = \frac{16}{20} - \frac{15}{20} = \frac{1}{20}$

$\frac{1}{20}$ of a tank = 7.5 litres, so a full tank holds $7.5 \times 20 = 150$ litres.

All Units 1 2 3

Practise... Fractions, decimals and rounding

G F E D C

Where possible use a calculator to check the answers to non-calculator questions.

1 Give each answer in its simplest form.

a What fraction of this shape is shaded?

b What fraction is not shaded?

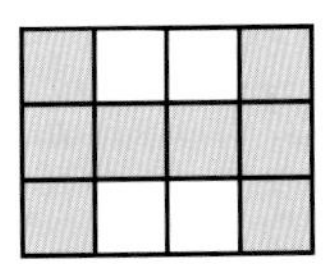

2 Simplify the following fractions:

a $\frac{5}{10}$ **b** $\frac{8}{10}$ **c** $\frac{12}{20}$ **d** $\frac{15}{50}$ **e** $\frac{75}{100}$

3 Work out

a $\frac{1}{5}$ of 365 **b** $\frac{2}{3}$ of 120 km **c** $\frac{3}{4}$ of £96 **d** $\frac{5}{8}$ of 856 kg

4 Arrange these fractions in order of size, starting with the smallest.

$\frac{7}{9}$ $\frac{5}{6}$ $\frac{2}{3}$ $\frac{3}{4}$ $\frac{7}{12}$

5 Put these numbers in order of size, starting with the smallest.

0.76 0.667 0.67 0.7 0.759 0.676

6 The population of a city is 82 795. Write this number:

a to the nearest thousand

b to the nearest hundred

c to the nearest ten.

7 **a** Work these out: **i** 18.4 + 7.92 **ii** 5.6 − 2.38

b Use the inverse operation to check each answer.

8 Work these out: **a** $\frac{3}{7} + \frac{2}{3}$ **b** $\frac{7}{10} - \frac{1}{2}$

9 **a** Estimate the value of 39.8 × 4.75

b Is your answer to part **a** an under-estimate or an over-estimate? Give a reason for your answer.

10 Use your calculator to work out $\frac{2.46 + 5.89}{3.92 - 1.75}$

a Give your answer as a decimal and write down all the figures in your calculator display.

b Round your answer to: **i** 1 decimal place **ii** 1 significant figure.

11 Write these fractions as decimals:

a $\frac{7}{10}$ **b** $\frac{4}{5}$ **c** $\frac{3}{4}$ **d** $\frac{13}{20}$ **e** $\frac{2}{3}$

12 Work these out: **a** $\frac{6}{7} \times \frac{3}{4}$ **b** $\frac{9}{10} \div \frac{3}{5}$

13 Write these decimals as fractions. Give each answer in its simplest form

a 0.6 **b** 0.25 **c** 0.85 **d** 0.56 **e** 0.375

14 At an audition 24 dancers out of 60 pass.
Give each fraction in its simplest form.

a What fraction of the dancers pass the audition?

b What fraction of the dancers fail the audition?

15 **a** Work these out: **i** 2.4×0.3 **ii** $7.56 \div 0.6$

b Use the inverse operation to check each answer.

16 Work these out.

a $2\frac{3}{5} + 4\frac{1}{2}$ **b** $5\frac{2}{3} - 1\frac{5}{6}$ **c** $3\frac{1}{5} \times 2\frac{3}{4}$ **d** $2\frac{5}{8} \div 3\frac{1}{2}$

17 Which of these numbers are the reciprocal of $\frac{4}{5}$?

a 0.45 **c** $1\frac{1}{5}$ **e** 1.25 **g** $1\frac{1}{4}$

b $\frac{5}{4}$ **d** 0.8 **f** 1.2

18 A shop usually sells rugs for one-third more than it pays for them.
In a sale it reduces the selling price by a third.
A shop assistant says the shop is making a loss on the rugs it sells in the sale.
Is the shop assistant right? Explain your answer.

19 Carl is buying some DVDs. Each DVD costs £7.95.

a How many DVDs can Carl buy for £35?

b How much money does he have left?

20 This is part of Delia's electricity bill.

Electricity Company	
Present reading - 8628 units	Last reading - 6953 units
Each unit costs 12.8p	

How much does she pay?

21 Sam uses $\frac{3}{4}$ of a tin of paint to paint a door.
Work out the **least** number of tins he needs to paint 6 identical doors.

22 **a** Find the fraction that lies halfway between 0.5 and 0.6.

b Find the decimal that lies halfway between $\frac{1}{5}$ and $\frac{1}{6}$.

23 Jack has £1.37 and Jill has 79 pence.

a How much must Jack give Jill so that they each end up with the same amount?

b Jack gives this amount to Jill using the smallest possible number of coins.
List the coins that Jack gives Jill.

Number

3 Percentages

Unit 1

4 Percentages

Unit 2

8 Percentages

Unit 3

4 Percentages and ratios

Key terms

Write down definitions for the following words. Check your answers in the glossary of your Student Book.

amount
balance
deposit
depreciation
discount
interest
percentage
principal
rate
Value Added Tax (VAT)

Revise... Key points

Fractions, decimals and percentages

All Units 1 2 3

To write a percentage as a fraction, write it with a denominator of 100 and simplify if possible.

For example, $45\% = \frac{45}{100} = \frac{9}{20}$ (numerator ÷ 5, denominator ÷ 5)

In Units 1 and 3 you can use a calculator. Make sure you can use the fraction key to simplify fractions.
In Unit 2 you need to be able to do everything without a calculator.

To write a percentage as a decimal, divide by 100.
For example, $45\% = \frac{45}{100} = 45 \div 100 = 0.45$

To write a decimal as a percentage, multiply by 100%.
For example, $0.7 = 0.7 \times 100\% = 70\%$

To write a fraction as a percentage multiply by 100%.
For example, $\frac{3}{5} \times 100\% = 60\%$

Or use equivalent fractions: $\frac{3}{5} = \frac{60}{100} = 60\%$ (numerator × 20, denominator × 20)

AQA *Examiner's tip*

When you are asked to compare fractions, decimals and percentages, it is usually easier to change them all to percentages.

Finding a percentage of a quantity, and increasing or decreasing an amount by a percentage

All Units 1 2 3

Unitary method – a useful method with a calculator and without a calculator if the numbers are easy enough.

- **Divide** the quantity **by 100** (to find 1%)
- then **multiply by the percentage** you need.
- For a percentage **increase**, **add** to the original amount.
  For a percentage **decrease**, **subtract** from the original amount.

Multiplier method – the most efficient method on a calculator.

- Write the percentage as a **multiplier** (by dividing it by 100).
  For example, 35% as a multiplier is $35 \div 100 = 0.35$
- then **multiply by the quantity**.

To increase or decrease a quantity by a percentage:

- Write the new quantity as a percentage of the original quantity.
- Convert this percentage to a **multiplier** (by dividing it by 100).
  For a 35% increase, the multiplier is $135 \div 100 = 1.35$
  For a 35% decrease, the multiplier is $65 \div 100 = 0.65$
- Multiply by the original quantity.

Using links with 10% – often the easiest method when you do not have a calculator (Unit 2).

- To find 10% of a quantity, divide it by 10 $\left(\text{because } \frac{10}{100} = \frac{1}{10}\right)$.
- Find other percentages using links with 10% (and/or 1%).
- For a percentage **increase**, **add** to the original amount.
  For a percentage **decrease**, **subtract** from the original amount.

Writing one quantity as a percentage of another

- Write both quantities in the same units.
- Divide the first quantity by the second to give a fraction or decimal.
- Write the fraction or decimal as a percentage.

AQA Examiner's tip

Remember you must use the same units when writing one quantity as a percentage of another.

Writing an increase or decrease as a percentage

All Units 1 2 3

- Find the increase or decrease.
- Divide the increase (or decrease) by the **original** amount to give a fraction or decimal.
- Write the fraction or decimal as a percentage.

$$\text{Percentage increase (or decrease)} = \frac{\text{increase (or decrease)}}{\textbf{original}\text{ amount}} \times 100\%$$

You can also use this method to find a percentage profit or loss.

$$\text{Percentage profit (or loss)} = \frac{\text{profit (or loss)}}{\textbf{original}\text{ (cost) price}} \times 100\%$$

Bump up your grade

For Grade C you should be able to write an increase or decrease as a percentage.

Example Comparing fractions and percentages

E

Greg and Cilla are on a 200-mile journey. Greg says they have gone about $\frac{1}{4}$ of the distance.
Cilla thinks they have gone about 40% of the distance. A signpost says they still have 130 miles to go.
Whose estimate is nearer, Greg's or Cilla's? Show how you decide.

Solution

There are different ways to work this out. You could compare miles:

Greg thinks they have gone $\frac{1}{4}$ of the way. $\frac{1}{4}$ of $200 = 200 \div 4 = 50$ (miles)
Cilla thinks they have gone 40% of the way.
10% of 200 = 20, so 40% of 200 = $20 \times 4 = 80$ (miles)

It is also possible to use fractions or decimals, but it is harder to compare them. Whole numbers or percentages are usually easier to compare.

In Unit 1 and Unit 3 you can work this out on your calculator: $200 \div 100 \times 40 = 80$ (miles)

The distance they have actually gone is $200 - 130 = 70$ miles
Cilla's estimate is nearer.

You could also answer the question by comparing percentages:

Cilla's estimate of the percentage they have travelled is 40%. Greg's estimate $= \frac{1}{4} = 25\%$
The distance they have actually travelled $= 200 - 130 = 70$ miles
The percentage they have travelled $= \frac{70}{200} = \frac{35}{100} = 35\%$
Cilla's estimate is nearer.

Example Increasing by a percentage

D

Sally earns £17 520 per year. She gets a pay rise of $2\frac{1}{2}$%.
How much does she earn per month after the increase?

Solution

Sally's pay rise = £17 520 ÷ 100 × 2.5
= £438 per year
Sally's new pay = £438 + £17 520
= £17 958
Sally's new pay each month = £17 958 ÷ 12
= £1496.5
= £1496.50

Alternatively, you could use the multiplier method here. Sally's new pay is 102.5% of her old pay.
The multiplier = 102.5 ÷ 100 = 1.025
and Sally's new pay per month
= 1.025 × £17 520 ÷ 12 = £1496.50
You can do this on a calculator in a single calculation.

AQA Examiner's tip

Remember to add a zero to the end of amounts of money where necessary to complete the pence.
Write down what you are doing, so that you get marks for the method even if you make a mistake on your calculator.

Example Decreasing by a percentage

C

The usual cost of insuring Aram's car is £450. His insurance company gives him a **discount** of 20% for not claiming in previous years. The discounted price is reduced by another 15% because Aram agrees to pay the first £100 of any claim he makes.

How much does Aram pay?

Bump up your grade

For a Grade C you need to be able to carry out more than one percentage change.

Solution

10% of £450 = £450 ÷ 10 = £45
20% of £450 = £45 × 2 = £90
The discounted price = £450 − £90 = £360

10% of £360 = £360 ÷ 10 = £36
5% of £360 = £36 ÷ 2 = £18
15% of £360 = £36 + £18 = £54
The amount Aram pays = £360 − £54 = £306

Example Writing one quantity as a percentage of another

D

At a school, 762 students are right-handed and 92 are left-handed.
What percentage of the students are left-handed? Give your answer correct to 1 decimal place.

Solution

The total number of students = 762 + 92 = 854

Percentage who are left-handed = $\frac{92}{854}$ × 100%
= 92 ÷ 854 × 100%
= 10.7728…
= 10.8% to 1 decimal place

You need to write the number of left-handed students as a percentage of the total number of students.

Example Writing an increase or decrease as a percentage

Unit 2

A shopkeeper buys notebooks in packs of 20 for £24 per pack.
She sells the notebooks for £1.80 each.
What is her percentage profit?

Solution

Cost price for 1 notebook = £24 ÷ 20 = £1.20
The profit on each notebook = £1.80 − £1.20 = 60 pence

AQA Examiner's tip

Remember to use the same units, and remember that the denominator should be the **original** amount.

Writing the profit as a fraction of the **original** (cost) price gives $\frac{60}{120} = \frac{1}{2}$
The percentage profit = 50%

An alternative method is to work out the % profit on the pack of 20 notebooks.
The shopkeeper sells 20 notebooks for 20 × £1.80 = 2 × £18 = £36
The profit = £36 − £24 = £12
Writing this as a fraction of the original (cost) price gives
$\frac{12}{24} = \frac{1}{2} = 50\%$

Bump up your grade

Being able to work out percentage increase or decrease will improve your chance of gaining a Grade C.

Practise... Percentages

All Units 1 2 3

1 For each shape, find:

a the percentage that is shaded

b the percentage that is not shaded.

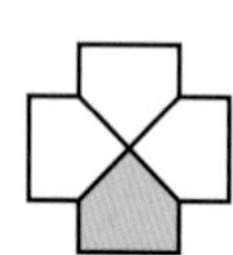
Shape A

Shape B

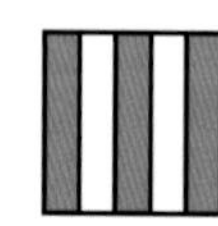
Shape C

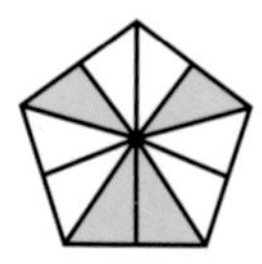
Shape D

2 Copy and complete the table. Write each fraction in its simplest form.

Decimal	Fraction	Percentage
0.3		
0.85		
0.375		
	$\frac{3}{4}$	
	$\frac{2}{5}$	
	$\frac{1}{3}$	
		7%
		28%
		2.5%
2.5		
	$1\frac{1}{4}$	
		420%

E

3 Write these in order of size, smallest first:

$\frac{4}{5}$, 0.54, 45%

4 Work these out:

a 25% of £560

b 70% of 1200 m

c 15% of 60 kg

d 35% of 40 litres

AQA Examiner's tip

It is always a good idea to **check** your answers. Use a different method if you can.

5 Calculate:

a 12% of £85

b 68% of 225 km

c 4% of £35.20

d 20% of £95.99

6 **a** Use multipliers to: **i** increase £4 by 20% **ii** reduce £4 by 20%

b Check your answers using a different method.

D

7 **a** Use multipliers to: **i** increase £135 by 6% **ii** reduce £135 by 6%

b Check your answers using a different method.

8 Rob and Simon are playing tennis.
Rob hits 36 aces out of 90 serves. Simon hits 42 aces out of 120 serves.
Does Rob or Simon have the greater percentage of aces from serves?
You **must** show your working.

9 The table shows the minerals and vitamins contained in a bowl of cereal and milk and the recommended daily amounts (RDA).

Copy and complete the table to show the percentage of each RDA that you get from the bowl of cereal and milk.

Give each answer to the nearest 1%.

Vitamin	Bowl contains	RDA	%
Iron	4.5 mg	14.0 mg	
Thiamine	0.5 mg	1.4 mg	
Riboflavin	0.7 mg	1.6 mg	
Niacin	5.9 mg	17.9 mg	

C

10 **a** In a sale the price of a tent goes down from £55 to £44.
What is the percentage reduction?

b The price of a magazine goes up from 96 pence to £1.20.
What is the percentage increase?

11 This year Jack's pay has risen from £17 500 to £18 200.
Jill's pay has risen from £24 000 to £24 840.
Whose pay rose by the greater percentage? You **must** show your working.

12 These scales show how much petrol there is in three identical petrol tanks.

a

b

0 10% 20% 30% 40% 50% 60% 70% 80% 90% 100%

c

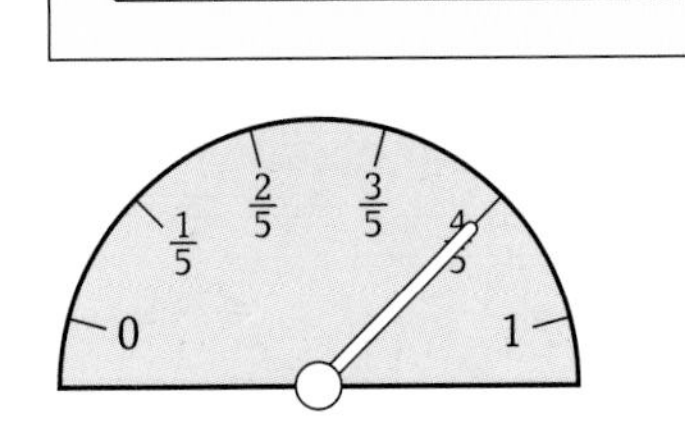

Which tank has the most petrol in it? Explain your answer.

13 A company buys a new delivery van for £27 500. The value of the van **depreciates** by 20% in the first year after it is bought. In the second year it depreciates by a further 16%.
What is the van worth when it is two years old?

Hint
Depreciates means decreases.

14 Julie invests £6500 at a fixed interest **rate** of 4% per year. At the end of the first year her **interest** is added to the **principal** and she leaves it all in the account for another year.

a What is the **amount** in Julie's account after interest is added at the end of the second year?

b What is the total interest earned over the two years?

15

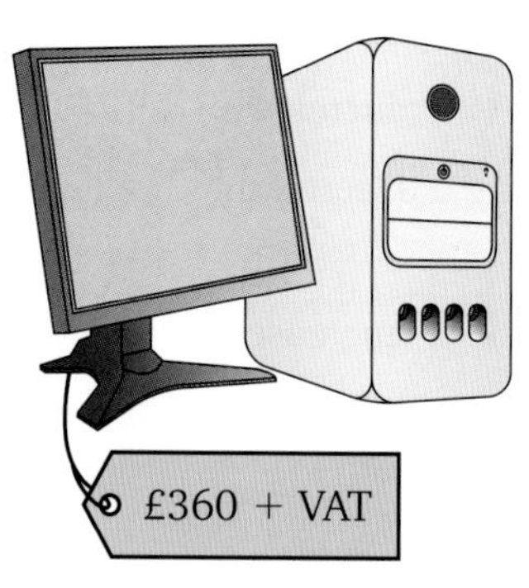

Work out the rise in price of a computer if the VAT rate goes up from $17\frac{1}{2}\%$ to 20%

16 Here is an advert for a summer holiday.

Fun in the sun

Dates	7 nights	14 nights
1 April – 31 May	£320	£525
1 June – 24 July	£290	£450
25July – 31 August	£395	£615

Notes
- Prices are for one adult (16 years and over).
- Children (less than 16 years) 75% of adult price.
- 10% discount if booked online.
- Pay 25% deposit when booking and the balance 8 weeks before departure.

Tanya books a seven-night holiday in May.
She books for herself, husband Alex and daughter Rosie (age 12 years).
She books the holiday online.

a Find the deposit that Tanya pays when she books the holiday.

b Tanya has 20 weeks to save money for the balance.
Workout how much Tanya needs to save each week.
Give your answer to the nearest pound.

17 Ruth, Sheila and Toby measure their heights.
Ruth's height is 85% of Toby's height. Sheila's height is $\frac{21}{25}$ of Toby's height.
Who is taller, Ruth or Sheila?
Explain your answer.

Number

4 Ratio and proportion

Key terms

Write down definitions for the following words. Check your answers in the glossary of your Student Book.

proportion

ratio

unitary method

unitary ratio

Revise... Key points

Finding and simplifying ratios

All Units 1 2 3

Ratios are used to compare two or more quantities.

To simplify a ratio:

- write all of the parts of the ratio in the same units, then omit the units
- divide each part of the ratio by the same number and repeat this until you have the smallest possible whole numbers.

In Unit 2 you will have to do this without a calculator.

In Units 1 and 3 you will be able to use the fraction key on your calculator to simplify ratios if you wish.

AQA *Examiner's tip*

It is very important to write the ratio in the correct order.
(3 : 10 is not the same as 10 : 3)

Links with fractions

All Units 1 2 3

There are a number of links between a ratio and fractions.

For example, when a brother and sister share some money in the ratio 2 : 3, this means that for every £2 the brother gets, the sister gets £3.

The brother gets $\frac{2}{3}$ of what the sister gets. The sister gets $\frac{3}{2} = 1\frac{1}{2}$ times what the brother gets.

The brother gets $\frac{2}{5}$ of the total amount. The sister gets $\frac{3}{5}$ of the total amount.

Unitary ratios

All Units 1 2 3

Sometimes it is useful to write a ratio in the form 1 : n or n : 1 where n may not be a whole number.

For example, in a school where there are 740 students and 50 teachers, the student : teacher ratio is 740 : 50
Dividing both sides by 50 gives 14.8 : 1

In Units 1 and 3 you will be able to do divisions like this on a calculator.

In Unit 2 you need to be able to work it out without a calculator. You could do this by dividing both sides of the ratio by 10, then 5.

This **unitary ratio** tells us that there are 14.8 students for every teacher at this school.
It is a useful ratio to use when comparing schools with different numbers of students and teachers.

Unitary ratios are also used on plans, maps and models. For example, the ratio 1 : 100 is often used on plans, the ratio 1 : 1250 is used in town planning, and the ratios 1 : 25 000 and 1 : 50 000 are used on Ordnance Survey maps.

To find the real distance between two places shown on a 1 : 50 000 map, multiply the distance on the map by 50 000.

Dividing quantities in a given ratio Unit 2

To divide something in a given ratio:

- add the numbers in the ratio to find the total number of parts
- divide the quantity by the total number of parts to find the amount for 1 part
- multiply by each number in the ratio to find the quantities required.

To check these quantities, add them together to make sure their total is the same as the quantity you started with.

AQA Examiner's tip

Always check that the total is correct after you have divided something in a ratio.

Ratio and proportion: the unitary method All Units 1 2 3

The unitary method is useful in many situations involving ratio and proportion.
It is based on finding the amount or cost of **one** unit.

Bump up your grade

To get a Grade C you should be able to use the **unitary method** to solve ratio and **proportion** problems.

For example, a recipe says you need 450 g of tomatoes to make 6 portions of tomato soup. You want to make 8 portions.
First work out the amount of tomatoes for **1 portion** of soup = 450 g ÷ 6 = 75 g
Then the amount of tomatoes for 8 portions = 75 g × 8 = 600 g

You can also use the unitary method to solve best-buy problems.

Remember you will be able to use a calculator to work this out in Units 1 and 3, but not in Unit 2.

Example Using ratios to find quantities Unit 2

C

Mrs Perfect uses this recipe to make a breakfast drink.

a Write the ratio of orange juice : pineapple juice : grapefruit juice in its simplest form.

b One morning Mrs Perfect makes $2\frac{1}{2}$ litres of breakfast drink. How much of each ingredient does she use?

Breakfast Fruit Juice

1 litre of orange juice
600 ml of pineapple juice
400 ml of grapefruit juice

Solution

a The ratio of orange juice, pineapple juice and grapefruit juice is 1000 ml : 600 ml : 400 ml = 5 : 3 : 2

Using 1 litre = 1000 ml. Dividing by 200 gives the simplest form.

AQA Examiner's tip

Remember the units must be the same before you simplify a ratio.

b $2\frac{1}{2}$ litres = 2500 ml
The total number of parts in the ratio is 5 + 3 + 2 = 10
This means that $\frac{5}{10}$ of the drink is orange juice, $\frac{3}{10}$ is pineapple juice and $\frac{2}{10}$ is grapefruit juice.
$\frac{1}{10}$ of 2500 ml = 2500 ml ÷ 10 = 250 ml
The amount of orange juice needed = $\frac{5}{10}$ of 2500 ml = 250 ml × 5 = 1250 ml
The amount of pineapple juice needed = $\frac{3}{10}$ of 2500 ml = 250 ml × 3 = 750 ml
The amount of grapefruit juice needed = $\frac{2}{10}$ of 2500 ml = 250 ml × 2 = 500 ml
To check this, work out 1250 ml + 750 ml + 500 ml = 2500 ml ✓

AQA Examiner's tip

Check that the amounts add up to the total.

Example Using the unitary method to solve a best-buy problem

Scott wants to buy some noodles. The supermarket sells these two bags of noodles.

a Which bag of noodles gives the better value for money?

b Why might Scott choose to buy the other bag?

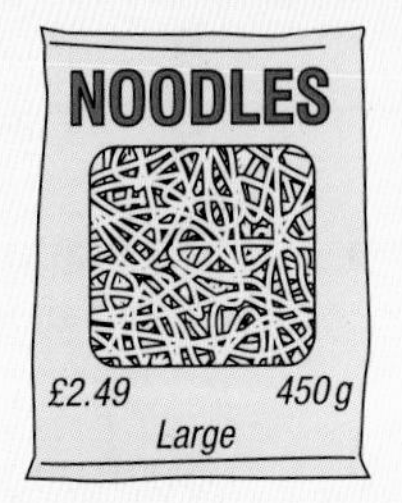

C

Solution

a Find the cost of **1 gram** in each bag.

Small: Cost of 300 g = 189 pence
Cost of 1 g = 189 ÷ 300 = 0.63 pence

Large: Cost of 450 g = 249 pence
Cost of 1 g = 249 ÷ 450 = 0.5533... pence

The cost of 1 gram of noodles is less in the large bag.
The large bag gives the better value for money.

b Scott might choose a small bag if he does not have enough money for the large bag or if he does not want any more than 300 g.

There are sometimes other ways to solve best-buy problems. In this case you could find the cost of 50 g:

Small bag (since 300 g = 6 × 50 g):
cost of 50 g = 189 ÷ 6 = 31.5 pence

Large bag (since 450 g = 9 × 50 g):
cost of 50 g = 249 ÷ 9 = 27.666... pence

This also shows that the large bag gives the best value for money.

Bump up your grade

To get a Grade C you should be able to solve best-buy problems.

Example Using ratio scales

The scale of a map is 1 : 250 000
On the map, the distance from a campsite to the nearest beach is 6 centimetres.
Work out the actual distance in kilometres.

C

Solution

The scale 1 : 250 000 means that the actual distance is 250 000 times the distance on the map.

Actual distance = 250 000 × 6 cm = 1 500 000 cm

Actual distance = 1 500 000 ÷ 100 ÷ 1000 km

Divide by 100 to convert centimetres to metres.
Then divide by 1000 to convert metres to kilometres.

The actual distance is 15 kilometres.

Practise... Ratio and proportion

D

1 Write each ratio in its simplest form.

a 18 : 12 **c** 250m : 1 km

b 100 : 160 **d** 90p : £1.20 : £2.10

Hint
Remember 1 km = 1000 m

2 **a** Divide £100 in the ratio 2 : 3 **c** Divide 42 miles in the ratio 1 : 2 : 3

b Divide £5400 in the ratio 3 : 7 **d** Divide 360 kg in the ratio 1 : 5 : 6

Hint
1 m = 100 cm

3 The simplest form of all of the following ratios is 1 : 4.
Find the missing numbers.

a 7 : ___ **b** ___ : 48 **c** 4.5 : ___ **d** ___ cm : 1 m

D

4 The ratio of girls to boys in a class is 5 : 4

a What fraction of the class are girls?

b What fraction of the class are boys?

c There are 27 students in the class. How many of these are girls and how many are boys?

C

5 The ratio of male to female workers in a factory is 7 : 13
What percentage of the workers are female?

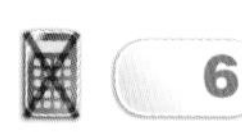
6 Six fruit scones cost £1.44.
How much does it cost for 10 fruit scones?

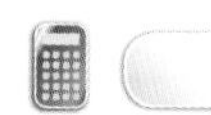
7 In a pie chart about how people travel to work, an angle of 60° represents 75 people.

a How many people does an angle of 40° represent?

b What angle represents 120 people?

8 Will, Kate and Neil are paid £64 between them for delivering leaflets.
They share this money in proportion to the number of leaflets they deliver.
Will delivers 48 leaflets, Kate delivers 60 leaflets and Neil delivers 84 leaflets.
Work out how much money each person gets.

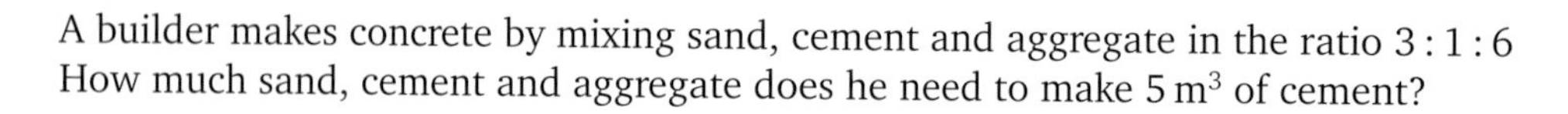
9 A builder makes concrete by mixing sand, cement and aggregate in the ratio 3 : 1 : 6
How much sand, cement and aggregate does he need to make 5 m^3 of cement?

10 An Ordnance Survey map has a scale of 1 : 50 000
The length of a road on the map is 17 cm. What is the actual length of the road in kilometres?

11 The table gives the number of teachers and students in four schools.

School	Number of teachers	Number of students
A	42	654
B	59	875
C	65	920
D	83	1246

a For each school, write the ratio of the number of teachers to the number of students in the form $1 : n$

b Which ratio do you think a parent would prefer? Explain your answer.

12 Lily wants to buy some rice.
The supermarket sells three different-sized bags of rice.

a Which size gives the best value for money?

b Give a reason why Lily might decide to buy a different size.

13 One 50 pence coin weighs 8 grams and contains copper and nickel in the ratio 3 : 1
In February 2010 the price of copper was £4.61 per kilogram and the price of nickel was £13.03 per kilogram.

How much, to the nearest penny, were the contents of a 50 pence coin worth in February 2010?

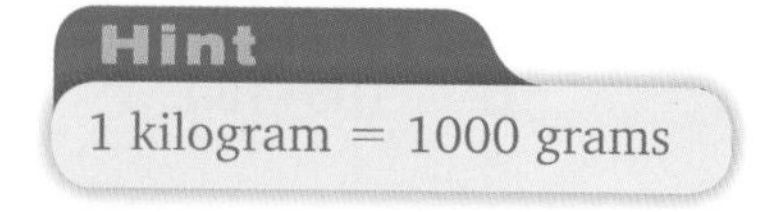

14 When Adel, Barry and Cathy set up a business, they contributed to the costs in the ratio 2 : 3 : 5
At the end of a year they invest 40% of the profits and share the rest in the ratio of their contributions.

Work out the percentage of the profits that each person receives.

Number

5 Indices

9 Indices

Key terms

Write down definitions for the following words. Check your answers in the glossary of your Student Book.

cube number
cube root
index
indices
power
square number
square root

Revise... Key points

Powers and roots (Unit 2)

Square numbers and square roots

The square numbers that you should know are:

$1^2 = \mathbf{1}$ $\quad 6^2 = \mathbf{36}$ $\quad 11^2 = \mathbf{121}$
$2^2 = \mathbf{4}$ $\quad 7^2 = \mathbf{49}$ $\quad 12^2 = \mathbf{144}$
$3^2 = \mathbf{9}$ $\quad 8^2 = \mathbf{64}$ $\quad 13^2 = \mathbf{169}$
$4^2 = \mathbf{16}$ $\quad 9^2 = \mathbf{81}$ $\quad 14^2 = \mathbf{196}$
$5^2 = \mathbf{25}$ $\quad 10^2 = \mathbf{100}$ $\quad 15^2 = \mathbf{225}$

You should also know the related square roots.
If you square a number, you always get a positive result.
For example, $5^2 = 5 \times 5 = 25$ and $(-5)^2 = -5 \times -5 = 25$ also.

AQA Examiner's tip
Remember: a negative number multiplied by a negative number gives a positive number.

So when you take the square root of 25 you get two answers, 5 and -5.

$\sqrt{25}$ means the positive square root of 25, so $\sqrt{25} = 5$.
The negative square root of 25 is $-\sqrt{25} = -5$

Cube numbers and cube roots

The cube numbers that you should know are:

$1^3 = \mathbf{1}$ $\quad 4^3 = \mathbf{64}$
$2^3 = \mathbf{8}$ $\quad 5^3 = \mathbf{125}$
$3^3 = \mathbf{27}$ $\quad 10^3 = \mathbf{1000}$

Remember 3^2 means $3 \times 3 = 9$ and 2^3 means $2 \times 2 \times 2 = 8$. Neither of them equals 6!

You should also know the related cube roots.

When you cube a negative number, you get a negative result.
For example, $(-5)^3 = -5 \times -5 \times -5 = -125$

When you take the cube root of a number, there is just one answer. $\sqrt[3]{125}$ means the cube root of 125.

$\sqrt[3]{125} = 5$ and $\sqrt[3]{-125} = -5$

AQA Examiner's tip
Remember that when you take the square root of a positive number there are two possible answers, but when you take the cube root there is just one possible answer.

Higher powers

The **index** or **power** tells you how many times to multiply the base number by itself.

This is the index or power

So $2^4 = 2 \times 2 \times 2 \times 2 = 16$

This is the base

and $10^6 = 10 \times 10 \times 10 \times 10 \times 10 \times 10 = 1\,000\,000$ (one million)

Rules of indices Unit 2

To multiply, add the indices.

So $4^6 \times 4^2 = 4^{6+2} = 4^8$ and in general $a^m \times a^n = a^{m+n}$

To divide, subtract the indices.

So $3^6 \div 3^5 = 3^{6-5} = 3^1 = 3$ and in general $b^m \div b^n = b^{m-n}$

To find a power raised to another power, multiply the indices.

So $(5^2)^3 = 5^{2\times 3} = 5^6$ and in general $(c^m)^n = c^{m\times n}$

because $(5^2)^3 = 5^2 \times 5^2 \times 5^2 = 5^6$

The rules of indices only work if the base number is the same in each term.
If not, you need to write each term as an ordinary number to multiply or divide them.

So $2^4 \times 5^2 = 16 \times 25 = 400$

Bump up your grade

You should be able to use these index rules for a Grade C.

Powers and roots on a calculator

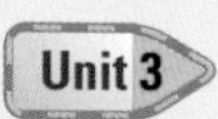

You may need to find a square, square root, cube, cube root or another power when you are using a formula.

Make sure that you can use your calculator to do this for difficult numbers. The keys you need usually look like this:

square

square root

cube

other powers

cube root

Unit 3 Chapter 3 (Working with symbols), Chapter 5 (Perimeter and area), Chapter 10 (Formulae), Chapter 11 (Area and volume), Chapter 18 (Pythagoras' theorem) all contain work that requires you to use these calculator keys.

Example Squares and square roots

Work these out:

a $(-7)^2$ **b** -7^2 **c** $\sqrt{\dfrac{100}{4}}$ **d** $\dfrac{\sqrt{100}}{4}$

Solution

a $(-7)^2 = -7 \times -7 = 49$

b -7^2 means $-(7 \times 7) = -49$ In **b** only the 7 is squared.

c $\sqrt{\dfrac{100}{4}} = \sqrt{25} = 5$ When the square root covers a calculation, the calculation must be done first.

d $\dfrac{\sqrt{100}}{4} = \dfrac{10}{4} = \dfrac{5}{2} = 2\frac{1}{2}$ or 2.5 In **d** the square root just covers 100, so find this before dividing by 4.

Example Cubes and cube roots

Which is bigger: 2^3 or $\sqrt[3]{64}$?

Solution

$2^3 = 8$

$\sqrt[3]{64} = 4$ since $4 \times 4 \times 4 = 64$

so 2^3 is bigger.

Example Rules of indices Unit 2

C

a Simplify the following giving your answer in index form:

i $3^4 \times 3^2$ **ii** $7^5 \div 7^3$ **iii** $(10^5)^3$ **iv** $p^5 \times p^3 \times p$ **v** $\frac{q^3 \times q^4}{q^2}$ **vi** $(r^2 \times r)^4$

b Kylie writes $\frac{4^6}{2^3 \times 2^2} = \frac{4^6}{4^5} = 4^1 = 4$

Terry writes $\frac{4^6}{2^3 \times 2^2} = \frac{4^6}{2^5} = 2^1 = 2$

i Explain the mistake each person has made.

ii Find the correct answer for $\frac{4^6}{2^3 \times 2^2}$

Solution

a **i** $3^4 \times 3^2 = 3^{4+2} = 3^6$

ii $7^5 \div 7^3 = 7^{5-3} = 7^2$

iii $(10^5)^3 = 10^{5 \times 3} = 10^{15}$

iv $p^5 \times p^3 \times p = p^{5+3+1} = p^9$ p is the same as p^1

v $\frac{q^3 \times q^4}{q^2} = q^{3+4-2} = q^5$

vi $(r^2 \times r)^4 = (r^3)^4 = r^{12}$

b **i** Kylie has multiplied the base numbers in the denominator. $2^3 \times 2^2 = 2^5$ **not** 4^5

Terry is correct up to $\frac{4^6}{2^5}$, but then he has divided the base numbers.

ii As the base numbers are different, you cannot use the rules of indices.

Each 4 is equal to 2×2

$$\frac{4^6}{2^5} = \frac{4 \times 4 \times 4 \times 4 \times 4 \times 4}{2 \times 2 \times 2 \times 2 \times 2} = \frac{\not{2} \times \not{2} \times \not{2} \times \not{2} \times \not{2} \times 2 \times 2 \times 2 \times 2 \times 2 \times 2 \times 2}{\not{2} \times \not{2} \times \not{2} \times \not{2} \times \not{2}} = 2 \times 2 \times 2 \times 2 \times 2 \times 2 \times 2 = 2^7 \text{ or } 128$$

Bump up your grade

It will help you to get a Grade C if you can use the rules of indices.

Practise... Indices 

G F E D C

F E

1 Find the value of:

a the square of 8 **c** -9^2 **e** $\sqrt{121}$ **g** 30^2 **i** 0.6^2

b 9^2 **d** $(-9)^2$ **f** $\sqrt{169}$ **h** $\sqrt{900}$ **j** $\sqrt{0.36}$

E

2 **a** Work out the sum of the two square numbers in this list:

27 32 49 56 66 72 81 96 125

b Work out the difference between the two cube numbers in this list:

27 32 49 56 66 72 81 96 125

3 Write down the value of:

a the cube of 10 **c** -2^3 **e** the cube root of 64 **g** $\sqrt[3]{27}$

b 2^3 **d** $(-2)^3$ **f** $\sqrt[3]{8}$ **h** $\sqrt[3]{-1000}$

4 Work these out:

a $3^2 - 2^3$ **b** $\sqrt{125 - 25}$ **c** $\sqrt[3]{125} - \sqrt{25}$ **d** $100^2 + \sqrt{100}$

5 Sharon says that 1 is the only number that is both a square number and also a cube number.
Is Sharon correct? Explain your answer.

6 Which of each pair is the greater?

a 9^2 or 92 **b** $\sqrt{36}$ or $\sqrt[3]{125}$ **c** $-\sqrt{25}$ or $\sqrt[3]{-27}$

7 Write down an approximate value for each of these.

a 6.9^2 **b** $(-3.2)^2$ **c** 4.93^3 **d** $(-0.995)^3$ **e** $\sqrt{83}$ **f** $\sqrt[3]{-7.95}$

Bump up your grade

For a Grade C you should be able to find the cube root of a negative number. You should also know that a positive number has a positive and negative square root.

8 Work out the value of each of the following:

a 10^4 **c** 2^5 **e** 5^4

b $(-10)^4$ **d** $(-2)^5$ **f** 4^5

9 Use the rules of indices to simplify the following. Give your answers in index form.

a $3^5 \times 3^4$ **d** $p^7 \times p^3$ **g** $\frac{5^8}{5^6}$ **j** $\frac{9^3 \times 9^2}{9}$

b $7^6 \div 7^2$ **e** $q^4 \div q^3$ **h** $(4^5 \times 4^2)^3$ **k** $\frac{t^8}{t^2 \times t^5}$

c $(6^3)^2$ **f** $(r^3)^5$ **i** $(8^7 \div 8)^4$ **l** $\left(\frac{n^6}{n^4}\right)^3$

10 Work these out. Give your answers as ordinary numbers.

a $2^4 \times 3^2$ **b** $10^3 \div 5^2$ **c** $\frac{9^3}{3^4}$ **d** $\frac{2^3 \times 5^2}{10^2}$ **e** $\frac{10^4}{2^3}$

11 Chloe says that $3^4 \times 3^3 = 3^{12}$

Dave says $3^4 \times 3^3 = 9^7$

a Explain the error that each person has made.

b What is the correct answer?

12 Julie is buying a new carpet.
The diagram shows its dimensions.
The carpet costs £10.99 per square metre and the underlay costs £4.99 per square metre. She pays £4 per square metre for having the carpet fitted.

Estimate the total cost.

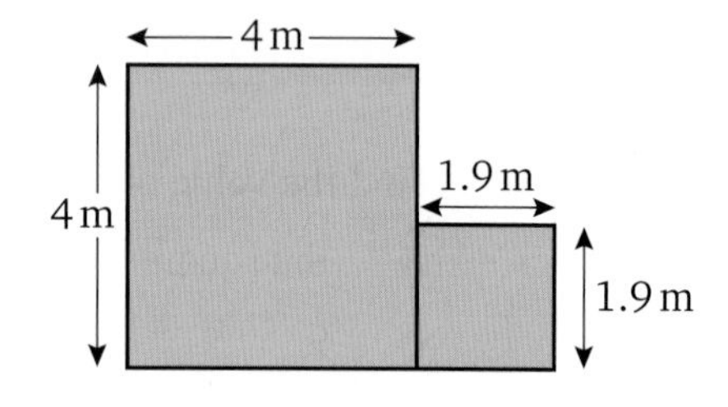

Link

Unit 3 Chapter 5 (Perimeter and area).

13 When you square any positive integer that has an 'end digit' of 1, the answer has the same 'end digit' (that is, the 'end digit' of the square is also 1).
For example, $11^2 = 121$, $21^2 = 441$, $31^2 = 961$ and so on.

a Explain why this occurs. **b** Find three other 'end digits' for which the same is true.

14 **a** Work out:

i $1 + 3$ **ii** $1 + 3 + 5$ **iii** $1 + 3 + 5 + 7$

b Find the sum of the first 100 odd numbers.

15 The number 81 can be written as $(3^2)^2$.
Find as many other ways as you can of writing 81 using indices.

AQA Examination-style questions

G F

1 Here are five digits: 2 5 8 1 7

a Use all of the digits to make the largest possible number. *(1 mark)*

b Use three of the digits to make an even number. *(1 mark)*

c Use some of the digits to make a number between 400 and 600. *(1 mark)*

d Use some of the digits to make two numbers that add up to 90. *(1 mark)*

e Use some of the digits to make two numbers with a difference of 25. *(1 mark)*

2 The cost of some ice creams at a shop are shown.

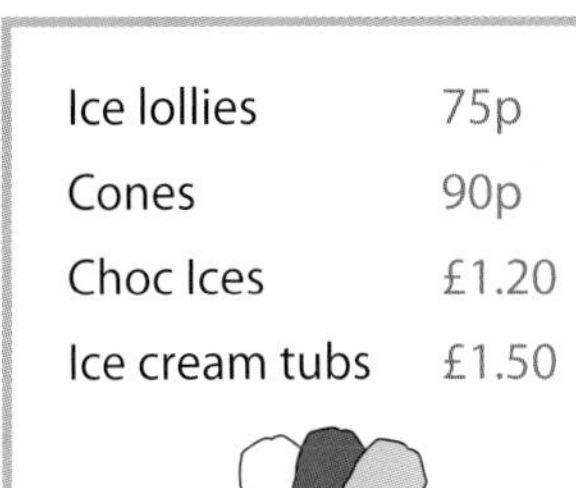
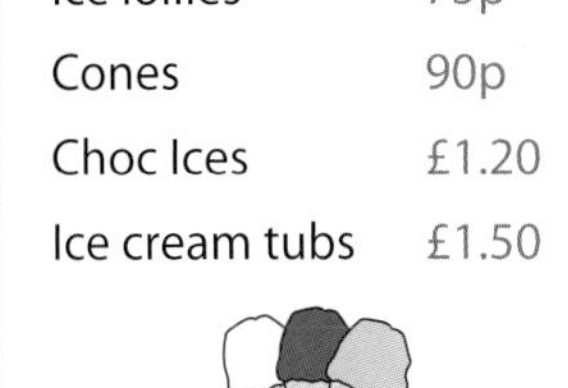

Ice lollies	75p
Cones	90p
Choc Ices	£1.20
Ice cream tubs	£1.50

a Tom buys two lollies and three cones.
How much change does he get from a £5 note? *(3 marks)*

b Nicola spends £2.70.
What could she have bought?
Give three possible answers. *(3 marks)*

c One day, the shop takes £36 for sales of lollies.
How many lollies are sold that day? *(3 marks)*

F

3 Here are three numbers. 36 49 57
Give a reason why each number could be the odd one out. *(3 marks)*

4 a Write $\frac{18}{81}$ as a fraction in its simplest form. *(1 mark)*

b Which fraction is smaller, $\frac{5}{7}$ or $\frac{5}{8}$?
Explain how you know. *(2 marks)*

c Given that $\frac{1}{6} = 0.166666\ldots$ write down $\frac{1}{6}$ of 1000.
Give your answer to one decimal place. *(2 marks)*

5 a How many twelfths are there in $\frac{2}{3}$?
You may use the grid to help you. *(1 mark)*

b What fraction of each grid is shaded?
Give your answers in the simplest form.

i

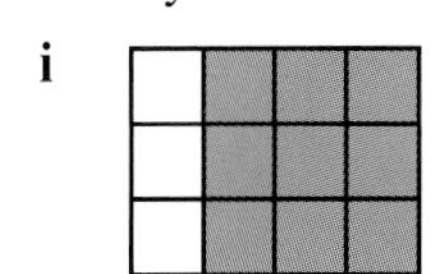

ii

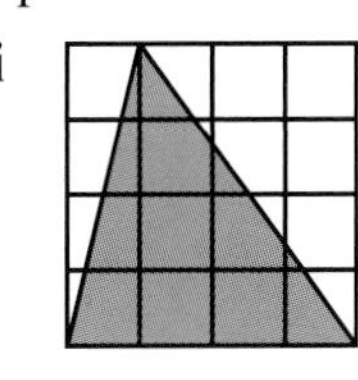

iii

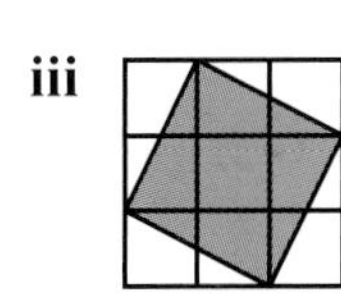

(3 marks)

Link

Unit 3 Chapter 5 (Perimeter and area)

E

6 A mobile phone contract costs £30 per month.

This includes: Option A: 250 free minutes of calls and 120 free texts, or
Option B: 90 free minutes of calls and unlimited free texts.

All extra calls cost 5p per minute. Extra texts cost 9p each.

On average, each month, Serena makes 450 minutes of calls and sends 200 texts.

Which option should she choose? You must show your working. *(5 marks)*

7 Mel goes shopping with £20.
She spends £1.30 on a coffee.
She needs £1.90 for her bus fare.
She sees this handbag.
Does she have enough money to buy it?
You **must** show your working. *(4 marks)*

8 Insert brackets on the left-hand side of each of the following to make them correct.

a $8 + 6 - 4 \div 2 = 5$ *(1 mark)*

b $9 - 2 \times 4 + 1 = 35$ *(2 marks)*

9 A gardener records the minimum temperature, in Celsius (°C) in his greenhouse each day.
The table gives the results for a week.

	Mon	Tues	Wed	Thurs	Fri	Sat	Sun
Temperature (°C)	1	0	−1	0	1	2	3

a Work out the difference between the maximum and minimum recorded temperatures. *(2 marks)*

b What fraction of the recorded temperatures are above 0 °C? *(2 marks)*

D

10 a Rachel earns £18 400 per year. She gets a pay rise of 3%.
How much is her pay rise? *(2 marks)*

b Steve earns £16 250 per year. He gets a pay rise of £520.
Write this as a percentage. *(2 marks)*

11 £2400 is shared in the ratio 7 : 3 : 2. Work out the largest share. *(3 marks)*

12 Three shops advertise the same bookcase.

Shop P
£40 deposit
plus 6 equal monthly payments of £15.

Shop Q
Usual price £165.
Special offer
$\frac{1}{3}$ off usual price.

Shop R
Usual price £140.
Special offer
20% off usual price.

At which shop is the bookcase cheapest? You must show your working. *(5 marks)*

13 Steve says, 'When you add two prime numbers together you always get an even number.'
Give an example to show that Steve is not correct. *(2 marks)*

14 a Work out:
i 0.8×0.2 *(1 mark)*
ii $2.4 \div 0.3$ *(1 mark)*
b Estimate the value of:
i 31.9×591 *(2 marks)*
ii $159.3 \div 21.95$ *(2 marks)*

15 a Write down all the factors of 28 and 42. *(2 marks)*
b What is the highest common factor of 28 and 42? *(1 mark)*

16 a Find both square roots of 4624. *(2 marks)*
b Calculate the cube of 1.7 *(1 mark)*
c Calculate $\frac{86.9 - 12.7}{2.4^2}$
i Write down your full calculator display. *(1 mark)*
ii Write your answer to one decimal place. *(1 mark)*
d Write down the reciprocal of 64. *(1 mark)*

17 a Work out the value of $\sqrt{36} \times \sqrt[3]{125}$ *(2 marks)*
b One of these statements is always true. Which one? Give a reason for your answer.
i The difference between the squares of two odd numbers is always odd.
ii The difference between the squares of two odd numbers is always even.
iii The difference between the squares of two odd numbers could be odd or even. *(2 marks)*

18 Rosie has two dogs, Bill and Ben.
Bill eats $\frac{1}{2}$ a tin of dog food each day. Ben eats $\frac{2}{3}$ of a tin of dog food each day.
What is the least number of tins of dog food that Rosie will need to feed both dogs for one week? *(4 marks)*

19 a Write 72 as a product of its prime factors. Give your answer in index form. *(2 marks)*
b Write these numbers as the product of their prime factors in index form:
i 144 *(1 mark)*
ii 720 *(1 mark)*

20 There are three bus routes across a town.
The number 52 bus leaves the station every 12 minutes, the number 42 bus leaves the station every 15 minutes and the number 86 bus leaves the station every 20 minutes.
The buses have just left the station together. How many times will they leave the station together in the next 12 hours? *(3 marks)*

21 The normal price of a 250 gram pack of butter is 90p.
There are two special offers.
Offer A 30% off the normal price. Offer B 30% extra butter. Price still 90p.
Which offer is the better value for money?
You **must** show your working. *(5 marks)*

22 In a sale the price of a computer is reduced from £600 to £480.
What is the percentage reduction? *(3 marks)*

23 Alice, Bernie and Carl share £700 between them.
Alice gets the largest amount of £320.
The ratio of Bernie's share to Carl's share is 1 : 4.
Work out how much Carl gets. *(3 marks)*

24 Sally wants to make some coconut cookies. She only has 100 g of butter.

How much coconut should she use? *(3 marks)*

25 **a** Work out $3^6 \times 3^4$
Give your answer as a power of 3. *(1 mark)*

b Work out $3^6 \div 3^4$
Give your answer as a whole number. *(1 mark)*

26 Work out the following:

a $2^3 \times 5^2$ *(2 marks)*

b x and y are prime numbers.
$xy^2 = 45$
Find x and y. *(2 marks)*

27 **a** Explain why $\sqrt[3]{100}$ has a value between 4 and 5. *(2 marks)*

b Which is smaller $-\sqrt{25}$ or $\sqrt[3]{-1000}$? Explain your answer. *(3 marks)*

28 **a** Use **two** whole numbers to write down a calculation with answer 3.5. *(1 mark)*

b Use **three** whole numbers to write down a calculation with answer 3.5. *(1 mark)*

c Work out 3.5×0.1 *(1 mark)*

Statistics

1 Collecting data

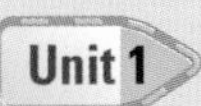

3 Collecting data

Key terms

Write down definitions for the following words. Check your answers in the glossary of your Student Book.

closed questions
continuous data
controlled experiment
data collection sheet
data logging
discrete data
frequency table
hypothesis
observation
observation sheet
open questions
pilot survey
population
primary data
qualitative data
quantitative data
questionnaire
raw data
sample
sample size
secondary data
survey
tally chart
two-way table

Key points

Data handling cycle Unit 1

The main reason for collecting data is to investigate a **hypothesis**. A hypothesis is a statement that you want to investigate (e.g. Do Year 10 girls take longer to get ready in the morning than Year 10 boys?). Specifying a hypothesis is the first stage of the data handling cycle.

Collecting the data that you need is the second stage of the cycle. The data that you first collect are called **raw data**. Raw data are data before they have been sorted.

Types of data Unit 1

Data can be primary or secondary.

Primary data are data that you collect yourself in order to investigate a hypothesis. For example, these may be data that you collected in a science experiment or a **questionnaire** you used on students.

Secondary data are information that has already been collected by someone else. For example, these may be data you have found on the internet or in a reference book.

Data can be qualitative or quantitative.

Qualitative data are non-numerical data. For example, hair colour or method of transport to get to work.

Quantitative data are data that take numerical values. For example, shoe size or height.

Quantitative data can be discrete or continuous.

Discrete data are data that can only take a set of fixed values. For example, shoe size or the number of days absent from work last year.

Continuous data are data that can take any value within a range. For example, the height of a Year 7 student or the speed of a free kick taken by a footballer.

Data collection methods Unit 1

A questionnaire is a common method for collecting data.

A questionnaire consists of a number of questions.

Questions may be **open** or **closed**.

Open questions allow for any response (answer) to be given (e.g. What did you have for breakfast?).

Closed questions have a set of responses that can be chosen. (See below for an example.)

All of the possible responses to a question are put into a response section.

How many singles have you downloaded in the last week?

0 ☐ 1–2 ☐ 3–5 ☐ 6–10 ☐ 11+ ☐

This is an example of a closed question.

This is the response section.

You may be asked in an exam to comment on a question or response section from a questionnaire.

The table below shows some questions and response sections and identifies the possible problems.

Question and response section	Problem
How old are you?	This question is too personal. If you want to ask this question you need to create a response section like this: 16 to 25, 26 to 35, 36 to 50, etc.
Don't you agree that fast food is unhealthy for you? Yes ☐ No ☐ Don't know ☐	This question is biased and it trying to encourage you to answer in a particular way.
How often do you exercise? Rarely ☐ Now and again ☐ Often ☐ Nearly every day ☐	The problem with this question is that there is no period of time indicated (e.g. in a week, month, etc.). The response section is also very vague and could be interpreted by people in different ways.
How many hours of TV did you watch last night? 0–1 ☐ 1–3 ☐ 3–5 ☐	Some of the response options are overlapping (e.g. where would you put 1 hour?). Also there is no option for over 5 hours.
Do you think more should be done to tackle global warming? Ye ☐ No ☐	This question requires a 'don't know' or 'no opinion' option box. This is for people who do not have an opinion on the matter or do not have enough information to be able to answer the question.

Raw data generated from a questionnaire are often organised in the form of a table.

This could be a **frequency table** or a two-way table.

Bump up your grade

To get a Grade C you need to be able to identify the different errors that can appear on questionnaires.

Organising data Unit 1

Once you have collected data, you need to organise it.

Two possible ways of organising the data are to use a **tally chart** or two-way table.

Goals scored	Tally	Frequency
0	𝍸 |	6
1	𝍸 𝍸 ||	12
2	||	2
3	|	1

A tally chart can be used to work out totals or frequencies.

(This chart shows the number of goals scored in 21 football games.)

	Coach A	Coach B	Coach C
Students	38	45	20
Teachers	4	5	2

A two-way table can be used to show more than one feature of the data at the same time.

(This table shows the number of students and teachers travelling on three different coaches.)

In an exam you could be asked to design a **data collection sheet** or an **observation sheet**.

A data collection sheet or an observation sheet are just ways of recording the data.

AQA *Examiner's tip*

If you are asked to create a data collection sheet then you should use either a tally chart or a two-way table.

Example Organising data Unit 1

The following data shows the marks obtained by 20 students in a test:

18	25	32	33	18	22	34	18	16	30
26	24	15	19	35	37	17	18	22	39

a Copy and complete the tally chart to show the information.

Marks	Tally	Frequency
15–19		
20–24		
25–29		
30–34		
35–39		

b What fraction of the class scored 25 marks or more?

c Here are some words that can be used to describe data.

Discrete Continuous Qualitative Quantitative

Select two words from this list that can be used to describe the data in this question.

Give a reason for each of your choices.

GF

Solution

a

Marks	Tally	Frequency
15–19	~~||||~~ |||	8
20–24	|||	3
25–29	||	2
30–34	||||	4
35–39	|||	3

You should always check that your frequency columns add up to the correct number.

We had 20 pieces of data so the frequency column should add up to this value.

$8 + 3 + 2 + 4 + 3 = 20$ This means our table contains 20 pieces of data.

AQA Examiner's tip

To avoid making mistakes try and tally the numbers one at a time. Once you have tallied a number, cross it off the original list. At the end check that the total of the frequency column is equal to the number of values in the original list.

b From the table you can see that $2 + 4 + 3 = 9$ people scored 25 marks or more.

There were 20 students in the class.

Therefore $\frac{9}{20}$ of the class scored more than 25 marks.

c The data are **discrete** as they can only take a set of fixed values (e.g. 1 mark, 2 marks, etc.).

The data are **quantitative** as they take numerical values.

Example Organising data Unit 1

D

A cinema screens a film three times a day.

The two-way table below shows the number of adults and children at each of the screenings.

	1pm screening	5pm screening	9pm screening
Adults	80	125	155
Children (under 16)	15	75	97

An adult ticket costs £6.50 and a child ticket costs £3.75.

The cinema has to pay £750 each time the film is screened.

Did the cinema make a profit from yesterday's screenings?

Show all your working.

Solution

The total number of adults = 80 + 125 + 155 = 360
The total ticket sales from adults = 360 × £6.50 = £2340

The total number of children = 15 + 75 + 97 = 187
The total ticket sales from children = 187 × £3.75 = £701.25

The total amount of ticket sales = £2340 + £701.25= £3041.25

Each screening costs £750. The cost of three screenings is £750 × 3 = £2250

The total amount of money taken from ticket sales is greater than the cost of screening the films.

This means that the cinema made a profit.

The amount of profit made was £3041.25 − £2250 = £791.25

Example Questionnaires Unit 1

C

Abdul and his family are travelling by train to London.

The customer service manager asks passengers to complete a questionnaire.

Here is one of the questions.

> How often do you travel with us and use our restaurant?
>
> Rarely ☐ Occasionally ☐ Often ☐ Very often ☐

a Write down one criticism of the question asked.

b Write down one criticism of the response section.

c How could the question and response section be improved?

Solution

a The question is made up of two questions.
It is possible that you travel with the company, but do not use the restaurant.

b The response section is very vague.
Different people may interpret the responses in different ways.

c The company could ask two separate questions, for example:

1 In the past month how many times have you travelled with us?

1–2 ☐ 3–5 ☐ 6–10 ☐ 11+ ☐

2 Have you used our restaurant services?

Yes ☐ No ☐

Practise... Collecting data 

1 A spinner has four coloured sections: red, blue, green and orange.

The spinner is spun 20 times and the results are shown below.

Green	Green	Red	Red	Orange
Green	Red	Blue	Orange	Red
Red	Red	Green	Blue	Red
Orange	Green	Red	Blue	Blue

a Complete a tally chart to show this information.

b Which colour occurs most often?

c Here are some words that can be used to describe data.

Discrete   Continuous   Qualitative   Quantitative

Select two words from this list that can be used to describe the data in this question.
Give a reason for each of your choices.

2 The amount of money spent by 18 customers at a supermarket checkout is shown below.

£7.25	£8.59	£18.60	£1.58	£13.40	£6.85
£8.00	£11.80	£4.47	£8.99	£6.55	£14.80
£1.56	£7.80	£3.58	£14.81	£26.50	£5.40

a Copy and complete the tally table below.

Money	Tally	Frequency
£0 $\leqslant x <$ £5		
£5 $\leqslant x <$ £10		
£10 $\leqslant x <$ £15		
£15 $\leqslant x <$ £20		
£20+		

Hint

The interval £5 $\leqslant x <$ £10 contains any values between £5 and £10. A value of £5 would be included here, but a value of £10 would not be included.

b What fraction of the customers spent less than £10?
Give your answer in its simplest form.

c Is this data an example of discrete data or continuous data?

3 Two classes study GCSE PE.

The two-way table below shows how many boys and girls are in each class.

	Boys	Girls
Class A	11	18
Class B	24	7

a How many girls study GCSE PE?

b How many students are in class A?

c What percentage of the girls are in class B?

d What fraction of class A are boys?

D

4 The table below shows the GCSE results in English and Maths of 250 students.

GCSE Maths Grade	GCSE English Grade							
	A*	A	B	C	D	E	F	G
A*	12	15	11	3	0	0	0	0
A	8	16	8	4	0	0	0	0
B	5	7	25	13	4	0	0	0
C	1	5	12	24	15	8	4	0
D	0	0	1	8	12	3	2	1
E	0	0	0	1	2	7	3	4
F	0	0	0	0	0	3	2	0
G	0	0	0	0	0	0	1	0

Fiona makes the following hypothesis:

'More students get a higher grade in English than maths.'

Do these results support this statement?

5 A local council wants to know people's opinions of their services.
They want to ask 200 people. Three possible ideas for conducting a **survey** are discussed.

Method 1	Method 2	Method 3
Visit the local shopping centre on Monday morning and ask 200 people.	Get the local phone book and ring 200 people at random.	Carry out a door-to-door survey in one particular area of town.

Write down a criticism for each method of collecting data.

C

6 In a town there are two supermarkets: Pricewise and Freshfoods.
Pricewise supermarket is carrying out a survey.
A member of staff stands in the entrance to their store and asks shoppers, at random, the following question.

Do you agree that our store provides better value for money than Freshfoods?

Strongly agree ☐ Agree ☐ Don't know ☐

a Write down one criticism of the question.

b Write down one criticism of the response section.

c Write down one criticism with the method of selection.

7 Mike is a driving instructor.
Mike believes that those students who have more lessons make fewer mistakes in their driving test.
Mike wishes to investigate this.
Put the following stages of the data handling cycle in the correct order:

Link

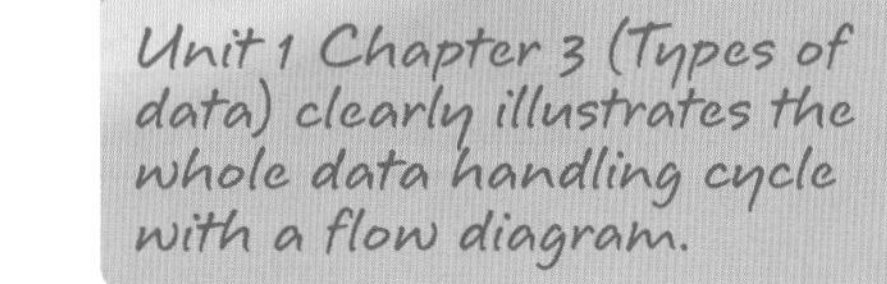

A Mike records the number of lessons taken and mistakes made by 30 of his students.

B Mike plots the data on a scatter diagram.

C Mike states the hypothesis, 'The more lessons a student has, the fewer mistakes they make'.

D Mike uses his diagram to conclude whether his hypothesis is true.

2 Statistical measures

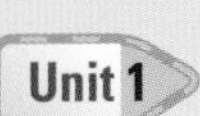

6 Statistical measures

Key terms

Write down definitions for the following words. Check your answers in the glossary of your Student Book.

average
class interval
continuous data
discrete data
frequency table
grouped data
mean
median
modal class
modal group
mode
range

Data handling cycle

Unit 1

The third stage of the data handling cycle is to process and represent the collected data. The fourth stage then involves interpreting the diagrams and calculations produced.

This chapter looks at processing the data and interpreting the calculations. Chapter 3.1 (Types of data) looks at using diagrams to represent and interpret the data.

Basic measures

Unit 1

Mean – Add up all the values and divide by the number of values you have.

Median – This is the middle value, when the values are listed in order. (If there are two numbers in the middle, the median is the mean of the two values.)

Mode – This is the value that appears the most.

Range – The difference between the highest and lowest values.

AQA *Examiner's tip*

Be careful when you are asked to give the range for a set of data. Your answer must be a single number (e.g. 6) and not a range of numbers (e.g. 3 to 9).

Frequency distributions

Unit 1

A frequency distribution shows how often individual values occur.

A frequency distribution is usually presented in the form of a **frequency table**.

Frequency distributions are normally used with **discrete data**.

Measures of **average**

When the data are presented in the form of a frequency table the **mean** is given by the formula

$$\text{mean} = \frac{\text{the total of (frequencies} \times \text{values)}}{\text{the total of frequencies}} = \frac{\Sigma fx}{\Sigma f}$$ where Σ means the sum of

Usually the x values are in the first column (or row) and the frequencies, f, are in the second column.

The **mode** is the value that appears the most. It is the value that corresponds to the highest frequency.

The **median** is the middle value when the values are listed in order.

The median value is $\frac{1}{2}(n + 1)$th value, where n is the total number of values (i.e. the total frequency).

Measures of spread

The **range** is the difference between the highest value and the lowest value in your table.

Grouped frequency distributions Unit 1

Grouped frequency distributions are usually used with **continuous data**.

Measures of average

When the values in the first column (or row) are grouped into **class intervals** then an estimate of the mean is calculated using the formula

$$\text{mean} = \frac{\text{the total of (frequencies} \times \text{midpoint)}}{\text{the total of frequencies}} = \frac{\Sigma fx}{\Sigma f}$$

This is only an estimate of the mean as we do not know the original exact values.

The **modal class** is the class or group that corresponds to the highest frequency.

The class interval that contains the **median** is the interval that contains the $\frac{1}{2}(n + 1)$th value.

Measures of spread

The range is estimated by finding the difference between the highest and lowest values in the table.
You would subtract the lowest value in the first class interval from the highest value in the last class interval.

Bump up your grade

To get a Grade C you need to be able to find an estimate of the mean of data in a grouped frequency distribution.

Example Basic measures Unit 1

F

a Here are six number cards.

Show that the median of these six cards is greater than the mean.

7	4	11	3	8	8

b A seventh card is added to the list.

The range of the cards increases by 2.

Work out the two possible values that the seventh card could be.

Solution

a To find the median of the cards, firstly put them in order.

3 4 7 8 8 11

Now find the middle number.

~~3~~ ~~4~~ 7 8 ~~8~~ ~~11~~

The middle number is the mean of 7 and 8.

The median is $\frac{7 + 8}{2} = 7.5$

To find the mean add up all of the values and divide by the number of values.

$$\frac{3 + 4 + 7 + 8 + 8 + 11}{6} = \frac{41}{6} = 6.83\ldots$$

$7.5 > 6.83\ldots$

This shows that the median is greater than the mean.

Hint

Here is a method to find the median. Cross off one number from the start and one number from the end of your list. Continue doing this until you are left with one number or two numbers.

b The range is the difference between the highest and lowest cards: $11 - 3 = 8$

When the seventh card is added the range increases by 2.

So the new card could be two more than the highest value, or two less than the lowest value.

The new card could be $11 + 2 = 13$, or the new card could be $3 - 2 = 1$

Example Frequency distributions Unit 1

D

Mr George teaches class 9R one lesson per week.

There are 32 students in the class.

Mr George records the number of students absent from his lesson each week over the whole school year.

The results are shown in the table.

Number of students absent	Number of weeks
0	5
1	14
2	12
3	4
4	2
5	1
6	1

Hint

Think of the first column as being the x values. Think of the second column as being the frequencies, f.

a Write down the modal number of students absent.

b Find the median number of students absent.

c Find the mean number of students absent.

d Use your answer to part **c** to calculate the mean number of students present.

e Find the range of students absent.

Solution

a The mode is the value that corresponds to the highest frequency.

The highest frequency is 14, which corresponds to the value of 1 student.

Therefore the modal number of students absent is 1 student.

AQA Examiner's tip

Be careful that you do not write down the frequency that corresponds to the mode as your answer.

b The median number of students absent corresponds to the $\frac{1}{2}(n + 1)$th value.

n is the total frequency $= 5 + 14 + 12 + 4 + 2 + 1 + 1 = 39$

So the median is the $\frac{1}{2}(39 + 1)$th $=$ 20th value

In order to find the 20th value you need to find the running totals.

Number of students absent, x	Number of weeks, f	Running total
0	5	5
1	14	$5 + 14 = 19$
2	12	$5 + 14 + 12 = 31$
3	4	$5 + 14 + 12 + 4 = 35$
4	2	$5 + 14 + 12 + 4 + 2 = 37$
5	1	$5 + 14 + 12 + 4 + 2 + 1 = 38$
6	1	$5 + 14 + 12 + 4 + 2 + 1 + 1 = 39$

The 20th value lies in here. This corresponds to 2 students absent. This is the median.

The median value is 2 students absent.

c Firstly add an extra column at the end of the table to contain the frequency × value ($f \times x$).

Number of students absent, x	Number of weeks, f	$f \times x$
0	5	$5 \times 0 = 0$
1	14	$14 \times 1 = 14$
2	12	$12 \times 2 = 24$
3	4	$4 \times 3 = 12$
4	2	$2 \times 4 = 8$
5	1	$1 \times 5 = 5$
6	1	$1 \times 6 = 6$
Totals	$\Sigma f = 39$	$\Sigma fx = 69$

AQA Examiner's tip

Make an extra column at the end of your table to put your calculations in. This will help the examiner to see your method.

The mean number of students absent per lesson is

$$\text{mean} = \frac{\text{the total of (frequencies} \times \text{values)}}{\text{the total of frequencies}} = \frac{\Sigma fx}{\Sigma f}$$

$$= \frac{\Sigma fx}{\Sigma f}$$

$$= \frac{69}{39}$$

$$= 1.7692\ldots$$

mean = 1.77 (to 2 d.p.)

d There are 32 students in the class and the mean number of students absent per lesson is 1.77.
The mean number of students present must be 32 – 1.77 = 30.23 students

e The range is the difference between the highest and lowest values.
The highest number of students absent is 6.
The lowest number of students absent is 0.
The range is 6 – 0 = 6 students.

Example Grouped frequency distributions 

C

The table below shows the prices of 300 selected houses in a town during 1999.

House price, x, (£000s)	Number of houses in 1999
$60 \leqslant x < 100$	78
$100 \leqslant x < 150$	127
$150 \leqslant x < 200$	49
$200 \leqslant x < 300$	27
$300 \leqslant x < 400$	19

a For the year 1999 calculate the class interval in which the median house price lies.

The mean house price of the same houses in 2009 was £200 000.

b Calculate an estimate of the percentage increase in the mean price of the houses between 1999 and 2009. Comment on the reliability of your answer.

Solution

a The median class interval is the one that contains the $\frac{1}{2}(n + 1)$th value.
n is the total frequency, which is 300 houses.
So the median is the $\frac{1}{2}(300 + 1)$th = 150.5th value.

In order to find the 105.5th value you need to find the running totals.

House price, x, (£000s)	Number of houses in 1999	Running total
$60 \leqslant x < 100$	78	78
$100 \leqslant x < 150$	127	$78 + 127 = 205$
$150 \leqslant x < 200$	49	$78 + 127 + 49 = 254$
$200 \leqslant x < 300$	27	$78 + 127 + 49 + 27 = 281$
$300 \leqslant x < 400$	19	$78 + 127 + 49 + 27 + 19 = 300$

The 150.5th value lies in here. This corresponds to the class interval $100 \leqslant x < 150$.

So the interval that contains the median is $100 \leqslant x < 150$.

b You first need to find an estimate of the mean house price in 1999.

In order to find the mean you need to calculate

$$\text{mean} = \frac{\text{the total of (frequencies} \times \text{midpoint)}}{\text{the total of frequencies}} = \frac{\Sigma fx}{\Sigma f}$$

Add two extra columns at the end of the table. In the first extra column you should list the set of midpoints of the class intervals.

Use the second column to calculate frequency × midpoint.

House price, x, (£000s)	f	Midpoint	Frequency × midpoint
$60 \leqslant x < 100$	78	80	$78 \times 80 = 6240$
$100 \leqslant x < 150$	127	125	$127 \times 125 = 15\,875$
$150 \leqslant x < 200$	49	175	$49 \times 175 = 8575$
$200 \leqslant x < 300$	27	250	$27 \times 250 = 6750$
$300 \leqslant x < 400$	19	350	$19 \times 350 = 6650$
Totals	$\Sigma f = 300$		$\Sigma fx = 44\,090$

The mean house price in 1999 is

$$\text{mean} = \frac{\text{the total of (frequencies} \times \text{midpoint)}}{\text{the total of frequencies}} = \frac{\Sigma fx}{\Sigma f} = \frac{44\,090}{300} = 146.9666\ldots$$

Since the house prices are given in £000s an estimate of the mean house price in 2000 is £146 967.

In 2009 the mean house price is £200 000.

The percentage increase is given by

$$\begin{aligned}\%\text{ increase} &= \frac{\text{increase}}{\text{original amount}} \times 100 \\ &= \frac{200\,000 - 146\,967}{146\,967} \times 100 \\ &= 36\% \text{ (to 2 s.f.)}\end{aligned}$$

This is only an estimate of the house price increase as you do not have the actual values.

Practise... Statistical measures

G

1 Georgia and Sarah are playing a game.
The game involves selecting five number cards.
Here are Sarah's cards:

4 1 7 9 7

Here are Georgia's cards:

2 8 3 5 9

a What is the mode of Sarah's cards?

b Explain why Georgia's cards do not have a mode.

c Which girl has the higher median?

F

2 The number of letters in the first nine words of a book are:

6 5 3 1 2 3 1 8 2

a Find the mean number of letters per word.

b Find the median number of letters per word.

c Find the range.

E

3 The mean of five cards is 6.
The range is twice the mean.
None of the cards are the same value.
Write down a possible set of cards.

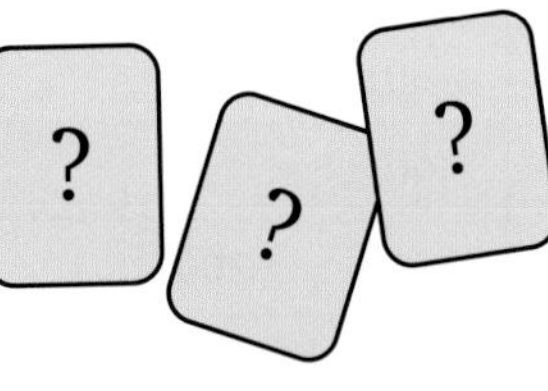

D

4 Richard is trying to find the mean of the following frequency distribution.

He says that the mean is 25.3...

a How do you know from the table that he is wrong?

b Find the mean of the data in the table.

x	Frequency
3	11
4	15
5	27
6	38

5 The number of letters received per day for 55 days is recorded.

Number of letters	Number of days
0	6
1	19
2	13
3	9
4	6
5	0
6	0
7	2

a Write down the modal number of letters received per day.

b Find the median number of letters received per day.

c Find the mean number of letters received per day.

d Find the range of letters received.

D

6 The heights of players in a basketball squad are shown below.

1.90 m	1.88 m	2.02 m	1.95 m	1.97 m
1.92 m	1.94 m	1.95 m	1.99 m	1.89 m

The heights of players in a football team are shown.

1.65 m	1.80 m	1.96 m	1.75 m	1.80 m	1.83 m
1.72 m	1.81 m	1.90 m	1.85 m	1.86 m	

Compare the heights of the players in the two teams.

C

7 Last year a European city decided to begin charging motorists to enter the city centre.
One year later the city authorities decided to look at the effect of the project.
The table below shows the number of occupants of all the cars entering the city between 8am and 9am on a typical work day.

Number of occupants	1	2	3	4	5	6
Number of cars before charging	658	275	86	25	4	1
Number of cars after charging	450	388	125	47	8	1

a Find the mean number of occupants per car before the charge was introduced.

b Find the mean number of occupants per car after the charge was introduced.

c Comment on the effect you think the charge has had.

8 One hundred people were asked how many text messages they had sent during the last week.

Text messages sent	Number of people
$0 < x \leqslant 10$	12
$10 < x \leqslant 20$	16
$20 < x \leqslant 40$	25
$40 < x \leqslant 100$	30
$100 < x \leqslant 200$	15
$200 < x \leqslant 500$	2

a State the modal class interval.

b Find the class interval that contains the median.

c Calculate an estimate for the mean number of text messages sent.

3 Representing data

7 Representing data
8 Scatter graphs

Key terms

Write down definitions for the following words. Check your answers in the glossary of your Student Book.

back-to-back stem-and-leaf diagram
bar chart
composite bar chart
continuous data
coordinates
correlation
dual bar chart
frequency diagram
frequency polygon
histogram
key
line graph
line of best fit
negative correlation
outlier
pictogram
pie chart
positive correlation
scatter graph
stem-and-leaf diagram
strength of correlation
type of correlation
zero or no correlation

Revise... Key points

Pictograms, bar charts and pie charts (Unit 1)

A **pictogram** is a diagram where a symbol is used to represent a certain number of items.

A **key** shows the number of items that the symbol represents.

A **bar chart** is a diagram where the length of each bar represents the frequency.

The bars on the chart are the same width.

Bar charts can be used to compare two sets of data at the same time.

A **dual bar chart** uses pairs of bars to compare data.

A **composite bar chart** uses single bars that are split into two or more parts.

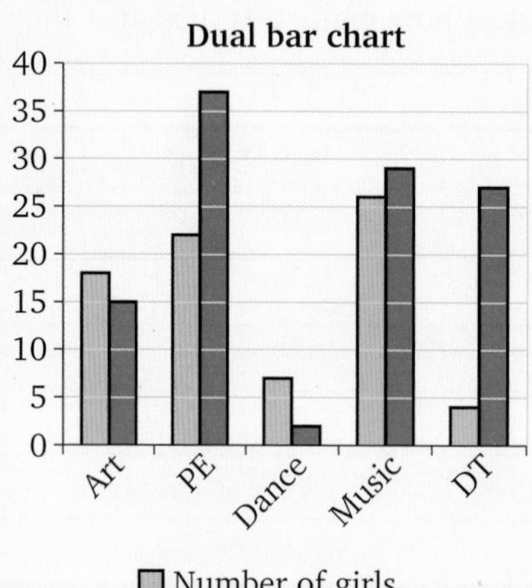

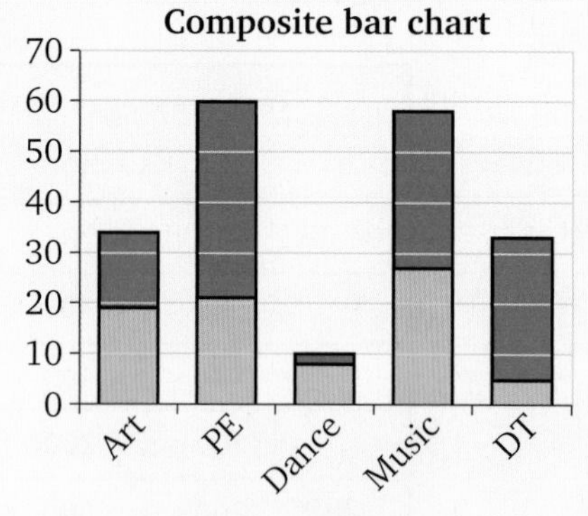

A **pie chart** is a diagram where the angle of each sector represents the frequency.

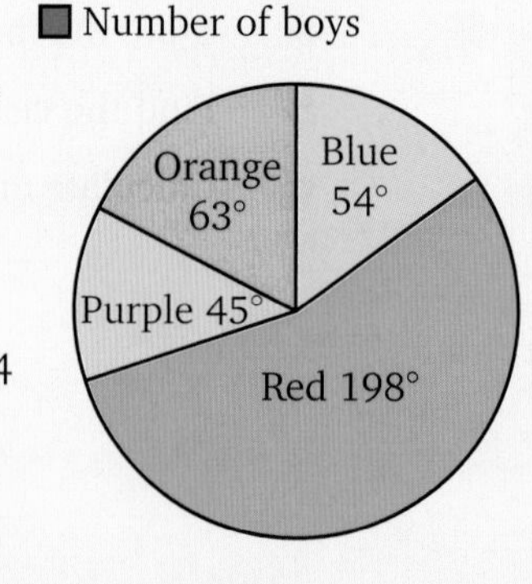

To find the angles for a pie chart

Step 1: Divide 360 by the total frequency.

Total frequency $= 12 + 44 + 10 + 14$
$= 80$

Now divide 360 by 80.

$360 \div 80 = 4.5$

Step 2: Multiply this answer by each frequency value to calculate the angles.

Colour	Frequency	Angle
Blue	12	$12 \times 4.5 = 54°$
Red	44	$44 \times 4.5 = 198°$
Purple	10	$10 \times 4.5 = 45°$
Orange	14	$14 \times 4.5 = 63°$

Check that all the angles sum to 360°.

AQA *Examiner's tip*

You must label your pie chart once you have drawn it. It must be clear which sector relates to which category. It is useful to label each sector with the angle and name.

Stem-and-leaf diagram Unit 1

A **stem-and-leaf diagram** is a way of representing data.

A stem-and-leaf diagram allows you to see the original data values.

Stem-and-leaf diagrams require a key so that you can interpret the values.

Stem-and-leaf diagram to show the heights of students on a minibus

Stem	Leaves
15	2
16	3 4 7 7
17	0 2 3 5 8
18	1 1
19	0

Key: 16 | 3 represents 163 cm

AQA **Examiner's tip**

In an exam you may be asked to draw a stem-and-leaf diagram. Ensure you draw an ordered stem-and-leaf diagram. This means you need to put the numbers in ascending order. Do not forget to include a key for your diagram.

Each leaf represents one value. So in this example, the heights of 13 students were measured.

Two data sets can be shown at the same time on a **back-to-back stem-and-leaf diagram**.

Link

See Learn 7.1 (Pictograms, bar charts and pie charts) for an example of a back-to-back stem-and-leaf diagram.

Line graphs, frequency polygons and histograms Unit 1

A **line graph** is a series of points joined with straight lines.

Line graphs show how data changes over a period of time.

A **frequency polygon** is an example of a **frequency diagram**.

It is used to display continuous grouped data.

For a frequency polygon you plot the frequency at the midpoint of the class interval.

Length of time taken	Number of people
$0 \leqslant t < 10$	6
$10 \leqslant t < 20$	22
$20 \leqslant t < 30$	31
$30 \leqslant t < 40$	15
$40 \leqslant t < 50$	3

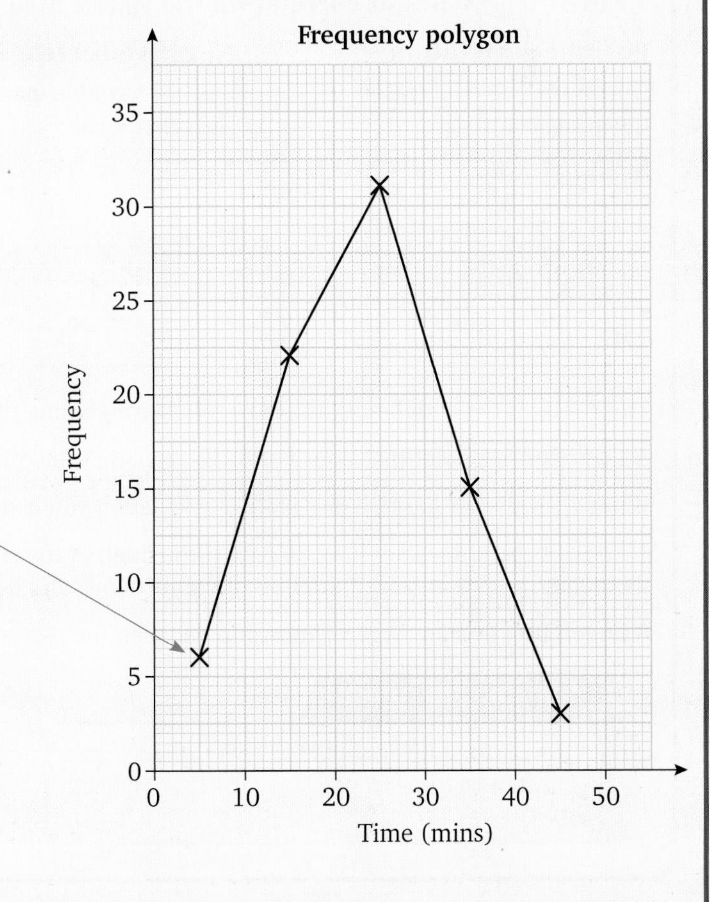

The first point is plotted at (5, 6).

5 is the midpoint of the interval and 6 is the frequency for that interval.

When you are asked to compare two frequency polygons you should make two comments.

The first comment should be about an average.

The second comment should be about the range.

Try and give your answers in the context of the question.

You could have also drawn a **histogram** to represent this data.

A histogram is a way of showing **continuous grouped data**.

The area of each bar is proportional to the frequency.

For a histogram with equal class widths, bars are drawn to the height of the frequency.

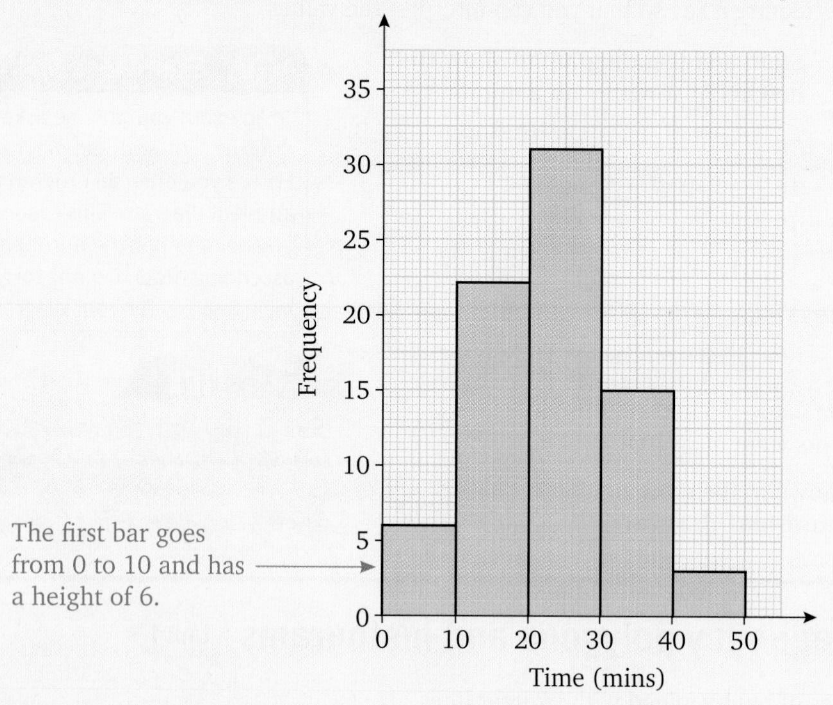

Scatter graphs Unit 1

Scatter graphs are used to show the relationship between two sets of data.

Correlation measures the relationship between two sets of data.

There are three **types of correlation** that you are likely to meet:

Positive correlation

(As one variable increases the other increases)

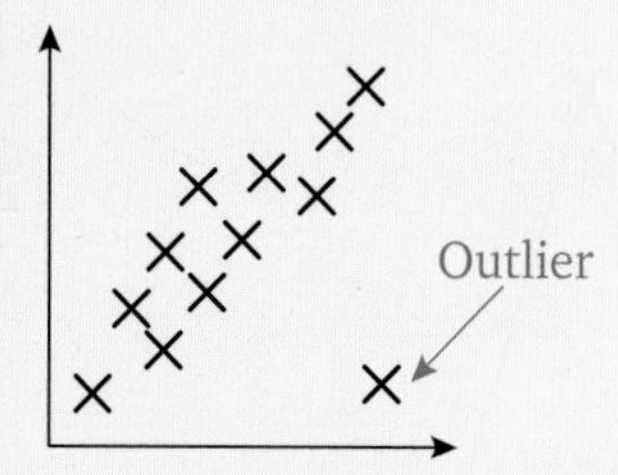

Negative correlation

(As one variable increases the other decreases)

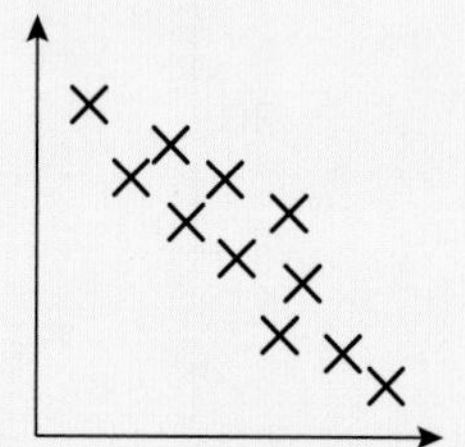

Zero or no correlation

(Points are scattered all over the diagram)

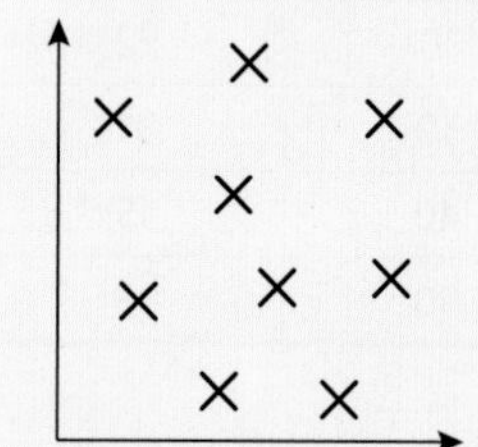

The stronger the correlation, the closer to a straight line the points lie.

An **outlier** is a point that does not fit the general trend of the data.

When a graph has positive or negative correlation you can draw a **line of best fit**.

A line of best fit does not have to pass through all of the points. The line should follow the general trend of the points, with a balanced number of points above and below the line.

A line of best fit shows the relationship between the two sets of data.

Bump up your grade

To get a Grade C you should be able to draw a line of best fit on a scatter graph and to use your line to estimate values.

Example Pictograms and bar charts Unit 1

G F

Alf sells new and used cars from Monday to Friday.

The pictogram shows the **total** number of cars he sold each day last week.

Monday

Tuesday

Wednesday

Thursday

Friday

Key: represents 2 cars

a How many cars did he sell on Tuesday?

Alf has a target to sell 25 cars in a week.

b Did Alf reach his target last week?

The composite bar chart shows the number of new and used cars Alf sold last week.

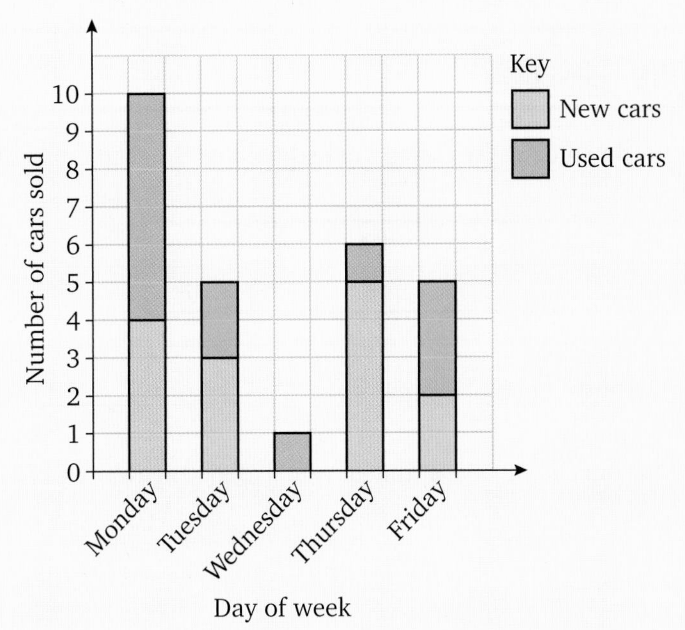

Alf gets paid a £50 bonus for each used car sold.
Alf gets paid a £120 bonus for each new car sold.

c What was Alf's total bonus last week?

Solution

a There are two full car symbols and one half car symbol on Tuesday.
One full symbol represents two cars, therefore $2 + 2 + 1 = 5$ cars were sold on Tuesday.

b On Monday Alf sold 10 cars.
On Tuesday he sold 5 cars.
On Wednesday he sold 1 car.
On Thursday he sold 6 cars and on Friday he sold 5 cars.
The total number of cars Alf sold was $10 + 5 + 1 + 6 + 5 = 27$ cars.
Alf met his target.

c Alf sold 14 new cars and 13 used cars.
For each new car sold he received a bonus of £120. So he received $14 \times £120 = £1680$
For each used car sold he received a bonus of £50. So he received $13 \times £50 = £650$
Alf's total bonus last week was $£1680 + £650 = £2330$

Example — Stem-and-leaf diagrams Unit 1

D

The stem-and-leaf diagram shows the number of marks that 25 students scored in an IQ test.

a Find the median score.

b Find the range of scores.

c What percentage of students scored more than 30 marks?

Stem	Leaf
1	5 6 6 7
2	2 3 4 5 5 8
3	0 1 4 7 9 9
4	2 5 6 7
5	1 2 3 9
6	0

Key: 2 | 3 represents a score of 23 marks

Solution

a The median score is the $\frac{1}{2}(n + 1)$th value. (Where n is the number of values.)
The median is the $\frac{1}{2}(25 + 1) = 13$th value.
The 13th value from the stem-and-leaf diagram is 34.

b The highest value is 60. The lowest value is 15.
The range is the difference between the highest and lowest values $= 60 - 15 = 35$ marks.

c Fourteen students scored more than 30 marks.
A total of 25 students took the IQ test.
The percentage of pupils who scored more than 30 marks is $\frac{14}{25} \times 100 = 56\%$

Example — Scatter graphs 

C

The table below shows the distance that eight students live away from school and the time it takes them to walk to school.

Distance (metres)	200	350	450	500	800	950	1100	1250
Time (minutes)	2	5	6	8	11	13	16	17

a Draw a scatter graph to show this information.

b Describe the relationship shown.

c Draw a line of best fit on your scatter graph.

d Use your line of best fit to estimate the time it will take a student who lives 600 m away to walk to school.

Solution

a

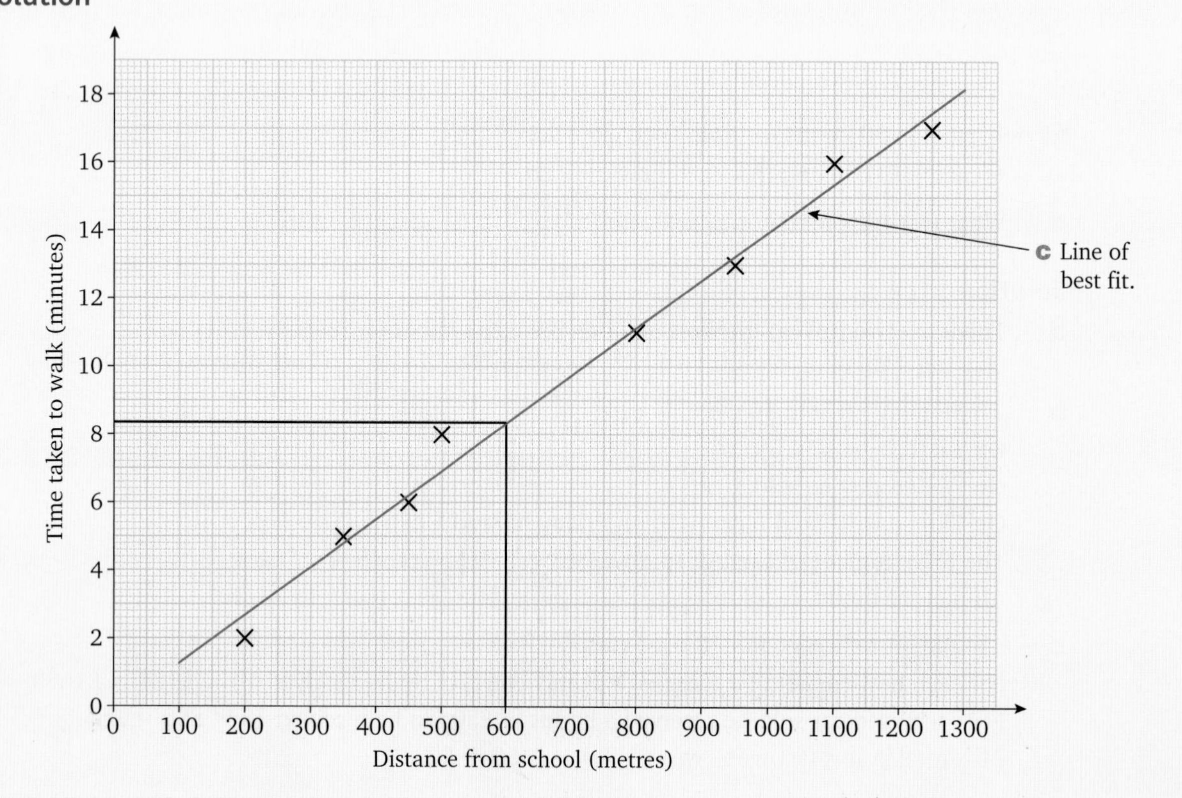

b The graph shows that the further a student lives from school the longer it takes them to walk to school. This is an example of positive correlation.

> **AQA Examiner's tip**
>
> When you are asked describe the relationship, you should try and explain what the graph is showing you. Do not just use the words positive and negative correlation. Try to give your answer in the context of the question.

d Using the line of best fit on the diagram, a student who lives 600 m away takes approximately 8.3 minutes to walk to school.

> **AQA Examiner's tip**
>
> When drawing a line of best fit, follow the general trend of the points and make sure you leave roughly the same number of points either side of the line.
> A line of best fit does not have to pass through the origin.

Practise... Representing data Unit 1

G F E D C

1 The number of customers per hour who visited a sweet shop was recorded.
The pictogram shows the results.

10am to 11am	⊙ ⊙ ⊙ ◖
11am to 12 noon	⊙ ⊙ ⊙ ⊙ ◖
12 noon to 1pm	⊙ ⊙ ⊙ ⊙ ⊙ ⊙ ⊙
1pm to 2pm	⊙ ⊙ ⊙ ⊙ ⊙ ◖
2pm to 3pm	⊙ ⊙
3pm to 4pm	

Key: ⊙ represents 4 customers

a How any customers were there between 11am and 12 noon?

b Between 3pm and 4pm there were 21 customers. Copy and complete the last row of the pictogram to show this information.

c 30% of the customers between 10am and 3pm were over 65.
How many customers were over 65?

G F

2 A town has one fire engine.
The bar chart shows the number of times the fire engine is called out each day during one week.

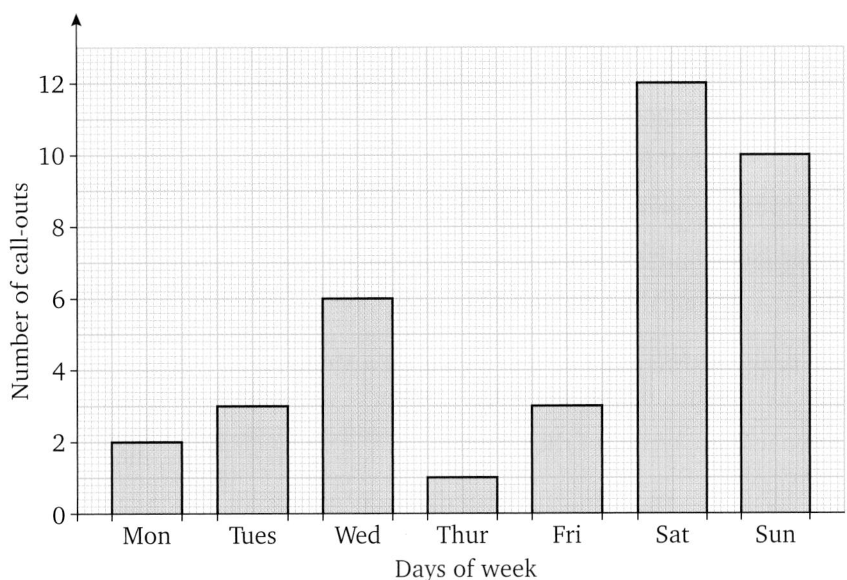

A newspaper article claims that the fire engine is twice as busy at the weekend as it is during the week.
Does the information in the chart support this claim?

F

E

3 Last year Evesthorpe Rovers played 16 football games.
The bar chart shows the number of goals scored in each of the matches.
The bar chart is incomplete.

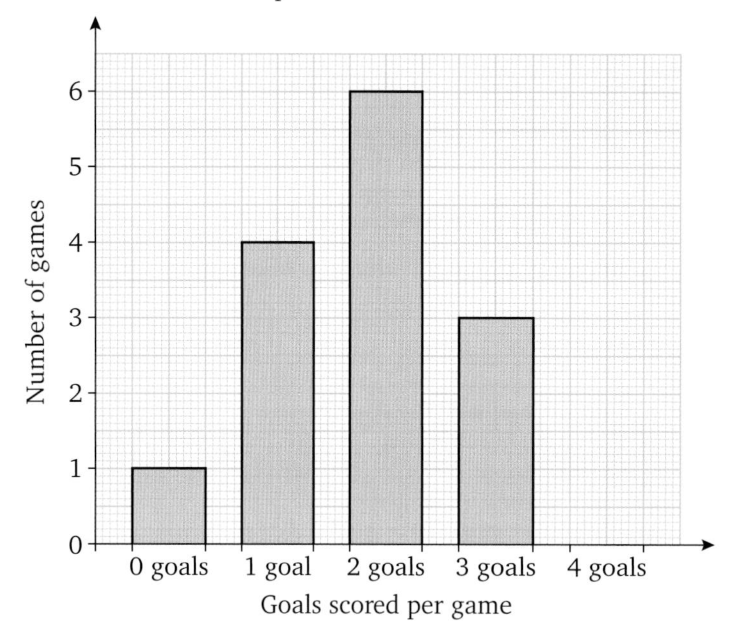

a Complete the bar chart to show the number of games in which 4 goals were scored.

b How many goals were scored in total in the 16 games?

4 120 students were asked how they travelled to school in the morning.
The table shows the results.

Method	Walk	Car	Bus	Other
Number of students	28	30	49	13

Draw a pie chart to represent this information.

5 The length of time that some people have to wait at a pedestrian crossing is shown in the stem-and-leaf diagram.

0	4 8 9
1	2 2 5 7 8
2	3 5 6
3	5 8
4	2 3 3

Key: 1 | 2 represents 12 seconds

a How many people had to wait at the pedestrian crossing?

b How many people were waiting longer than 30 seconds?

c Find the median length of time that a person had to wait.

d Find the range of time that people had to wait.

D C

6 A sports centre runs two different exercise classes on a Monday.
The table below shows the distribution of ages at one of the classes.

a Draw a frequency polygon for these data.

Age (years)	Frequency
20 up to 30	5
30 up to 40	11
40 up to 50	24
50 up to 60	36
60 up to 70	8

The frequency polygon shows the distribution of ages at the second class.

b Which class is more popular?

c Compare the ages of people who attend the two classes.

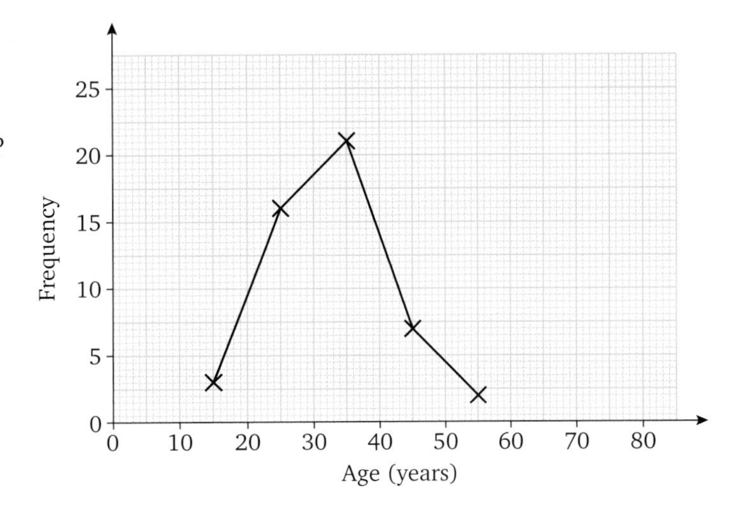

D C

7 The line graph shows the price of a train ticket over the last 10 years.

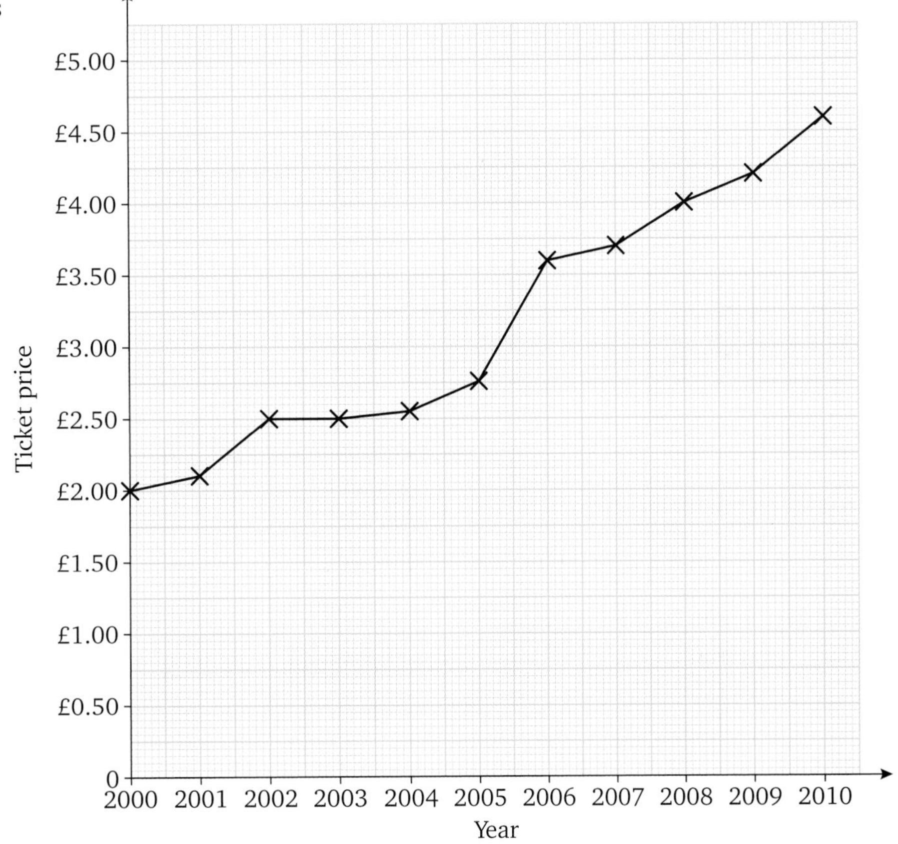

a What was the price of the train ticket in 2005?

b What was the price of the train ticket in 2001?

c Between which two years was the greatest increase in price? Calculate the increase in price between these two years.

d Jimmy says that the ticket price has increased by 120% over the last 10 years. Is Jimmy correct?

8 In a science experiment the amount of salt dissolving, in milligrams, was measured every one minute.
The results are shown in the table below.

C

Time (minutes)	1	2	3	4	5	6	7	8	9	10
Amount of salt remaining (mg)	40	38	35	34	32	30	29	28	25	25

a Plot this information on a scatter diagram and draw a line of best fit.

b Use your line of best fit to estimate how many milligrams of salt will be left after 14 minutes.
How reliable is your estimate? Explain your answer.

4 Probability

Unit 1

9 Probability

Key terms

Write down definitions for the following words. Check your answers in the glossary of your Student Book.

biased
certain
evens
event
experimental probability
fair
impossible
likely
mutually exclusive events
outcome
probability
probability scale
relative frequency
sample space diagram
theoretical probability
trial
two-way table
unlikely

Revise... Key points

Describing probability Unit 1

Probability is how **likely** an **event** is to happen.

The **probability scale** ranges from 0 to 1.

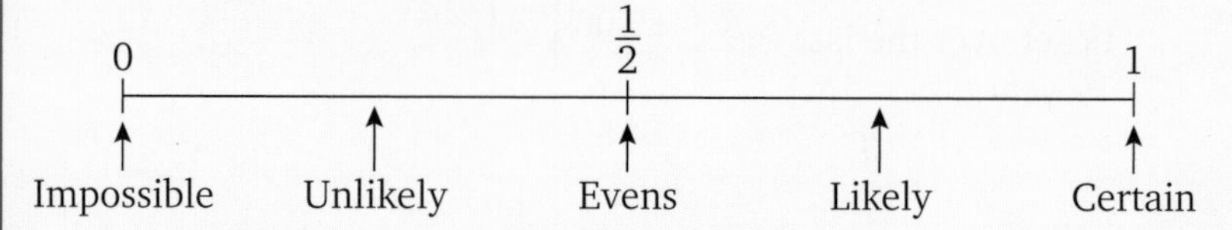

An **outcome** with a probability of 0 is **impossible**.

An outcome with a probability of 1 is **certain**.

All other probability values lie between 0 and 1.

Probability of an event happening $= \frac{\text{number of outcomes for that event}}{\text{total number of possible outcomes}}$

AQA Examiner's tip

Try to always use fractions or decimals to write probabilities. Never write a probability as a ratio. You will not get any marks for doing that in the exam.

Combining events Unit 1

Sometimes an event can come from trials with two parts (e.g. throwing a dice and spinning a spinner or throwing a dice twice etc.).

To show all possible outcomes you should use a **sample space diagram**.

A sample space diagram looks like a **two-way table**.

You can then use your table to work out probabilities.

1st dice score (columns); 2nd dice score (rows)

+	1	2	3	4
1	2	3	4	5
2	3	4	5	6
3	4	5	6	7
4	5	6	7	8

AQA Examiner's tip

Only cancel down a fraction if the exam question asks you to give your answer in simplest form.

Above is an example of a two-way table when two fair four-sided dice are thrown and their scores are added together.

A two-way table makes it easy to calculate probabilities, e.g. the probability of getting a total of 4 is $\frac{3}{16}$ since there are three totals of 4 out of 16 possible outcomes.

Mutually exclusive events Unit 1

Mutually exclusive events are events that cannot happen at the same time. For example, getting a two and getting an odd number when a dice is rolled once.

If the mutually exclusive events listed cover all of the possibilities, then the sum of their probabilities must equal 1.

Relative frequency Unit 1

Relative frequency is also known as **experimental probability**.

The relative frequency of an event is the probability of the event happening based on the results of an experiment.

$$\text{Relative frequency of an event} = \frac{\text{number of times an event has happened}}{\text{total number of trials}}$$

Relative frequency can also be based on past experience.

You can use results of an experiment to estimate the probability of an event.

The more times you carry out an experiment, the more reliable your probability estimate will be.

Bump up your grade

In order to get a Grade C you need to be able to calculate relative frequencies and plot them on a graph.

To work out the expected number of times a particular event will happen, you should multiply the probability by the number of times you are going to carry out the experiment.

Example Describing probability Unit 1

F

A box contains some coloured cubes.

There are 5 green cubes, 3 pink cubes and 2 blue cubes in the box.

I remove one cube at random from the box.

a What is the probability I remove a pink cube?

b What is the probability I remove a blue or green cube?

A second box contains 1 green cube, 1 pink cube and 1 blue cube.
I also remove one cube at random from the second box.

c Which box has a greater probability of having a pink cube removed? Explain your answer.

Solution

a There are 3 pink cubes in the box.

There are 10 cubes in the box in total.

We use the following for probability:

$$\text{Probability of an event happening} = \frac{\text{number of outcomes for that event}}{\text{total number of possible outcomes}}$$

The event is removing a pink cube and there are 3 pink cubes in the box.

The total number of possible outcomes is the total number of cubes.

Therefore:

$$\text{The probability of removing a pink cube is } \frac{\text{total number of pink cubes}}{\text{total number of cubes}} = \frac{3}{10} = 0.3$$

b The probability of removing a blue or green cube is $\frac{\text{total number of blue or green cubes}}{\text{total number of cubes}} = \frac{7}{10} = 0.7$

c The probability of removing a pink cube from the first box is $\frac{3}{10} = 0.3$

The probability of removing a pink cube from the second box is $\frac{1}{3} = 0.333333\ldots$

Therefore there is a greater probability you will remove a pink cube from the second box.

Example Mutually exclusive events Unit 1

D

A game involves choosing a card at random from 250 cards that are laid face down.

All of the cards are numbered 1, 2, 3 or 4.

The probability of choosing a particular numbered card is shown in the table below.

Card number	1	2	3	4
Probability	0.2	0.12		

There are twice as many cards numbered 3 as there are numbered 2.

Use all of this information to work out how many cards are numbered 4.

Solution

There are twice as many cards numbered 3 as there are numbered 2.

Therefore the probability of getting a 3 must be twice the probability of getting a 2.

Probability of getting a card numbered 3 = $0.12 \times 2 = 0.24$

The total probability in the table must equal 1 (as the events are mutually exclusive and they are all listed).

So the probability of getting a card numbered 4 must be

$1 - (0.2 + 0.12 + 0.24) = 0.44$

To work out the number of cards numbered 4 you need to multiply the probability by the total number of cards.

So the number of cards numbered 4 is $0.44 \times 250 = 110$ cards.

Example Relative frequency Unit 1

C

Daniel is carrying out an experiment to test whether a coin is **biased**.

He throws the coin 10 times and obtains 7 heads.

a Do these results suggest the coin is biased?
Explain your answer.

Daniel continues to throw the coin.
After every 10 throws he records the total number of heads obtained so far.
The results are shown in the table below.

Number of throws	10	20	30	40	50	60	70	80	90	100
Total number of heads	7	16	25	29	35	44	52	59	66	74

b Plot the results on a relative frequency graph.

In fact, the coin is biased.

c Write down an estimate of the probability of the coin landing on heads.

Solution

a No. Although the coin lands on heads more times than expected, Daniel has only thrown the coin 10 times. It is not possible to tell if the coin is biased from such a small number of trials.
Daniel needs to throw the coin more times.

b You first need to work out the relative frequencies.

For 10 throws the relative frequency is $\frac{7}{10} = 0.7$

For 20 throws the relative frequency is $\frac{16}{20} = 0.8$

You can work out the other relative frequencies in the same way.

Number of throws	10	20	30	40	50	60	70	80	90	100
Total number of heads	7	16	25	29	35	44	52	59	66	74
Relative frequency	$\frac{7}{10} = 0.7$	$\frac{16}{20} = 0.8$	$\frac{25}{30} = 0.83$	$\frac{29}{40} = 0.73$	$\frac{35}{50} = 0.7$	$\frac{44}{60} = 0.73$	$\frac{52}{70} = 0.74$	$\frac{59}{80} = 0.74$	$\frac{66}{90} = 0.73$	$\frac{74}{100} = 0.74$

You can now plot these points on a relative frequency graph.

Relative frequency of throwing heads

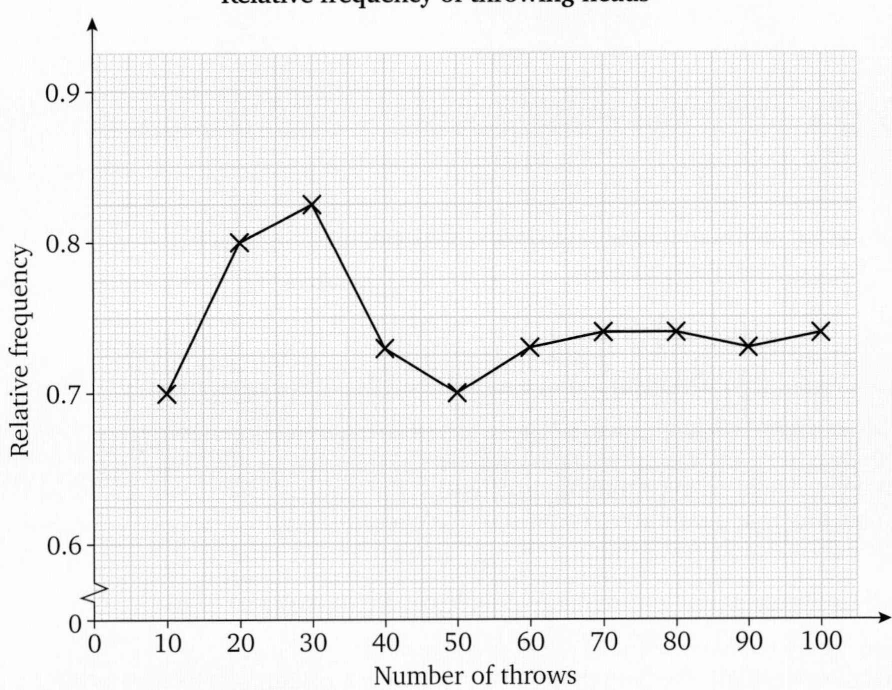

AQA Examiner's tip

Take care with the scale on any graphs. Before you plot any points you should check that you understand the scale being used on each axis.

c The best estimate of the probability of the coin landing on heads is 0.74.
It is the best estimate because it is the one that is based on most trials.

AQA Examiner's tip

Remember: the more times you repeat an experiment, the more reliable the probability.

Practise... Probability Unit 1

A jar contains some sweets.
4 of the sweets are strawberry, 2 are orange and 2 are mint.
Sonia chooses one sweet from the jar at random.
Sonia does not like mint sweets.

What is the probability that she chooses a sweet she likes?

F

E

2 A **fair** three-sided spinner is labelled 3, 4 and 5.
The spinner is spun once and a fair six-sided dice is rolled once.
The number that the spinner lands on and the dice score are added together.

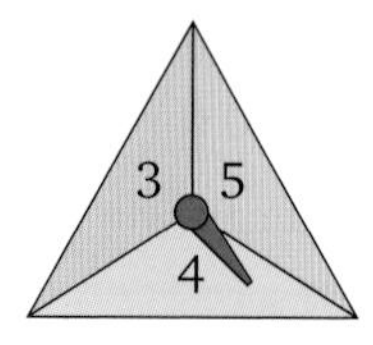

a Copy and complete the table to show all of the possible total scores.

Dice score

Spinner score	+	1	2	3	4	5	6
	3	4	5	6			
	4						
	5						

b Find the probability that the total score is greater than 6.

c Find the probability of getting an even total score.

3 Five numbered cards are placed face down on a desk.
Four of the cards are shown below.

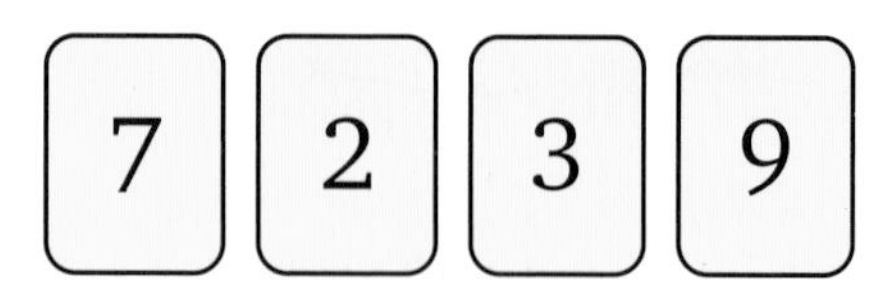

The probability that an even number is chosen is $\frac{2}{5}$.
The probability that a number greater than 5 is chosen is $\frac{3}{5}$.
Use this information to work out a possible value of the last card.

D

4 A spinner has 4 coloured sections – red, green, orange and blue.
The table below shows the probability of spinning each colour.

Colour	Red	Green	Orange	Blue
Probability	0.3	0.4	0.2	

a What is the probability of spinning red or green?

b What is the probability of spinning blue?

c What is the probability of not spinning red?

d I spin the spinner 120 times. How many times do I expect to spin red?

5 James, Andy and Dominic are playing card games.
The probability of each of them winning is shown in the table.

Player	James	Andy	Dominic
Probability	0.25	0.15	0.6

One evening they play 60 games.
How many more games should Dominic win than Andy?

C

6 A dice that is believed to be fair is rolled 30 times.

a How many times do you expect each number to appear?

Amanda decides to test whether the dice is fair.
She rolls the dice 30 times. The results are shown below.

2	2	3	3	5	3	3	6	1	3
5	3	3	3	3	2	4	1	5	4
6	2	3	3	2	3	1	4	3	3

b Copy and complete the relative frequency table below.

Score	1	2	3	4	5	6
Relative frequency						

c Do you think the dice is fair? Give reasons for your answer.

7 Julia is shooting arrows at an archery target.
Each time she aims for the centre.
She records the number of times she hits the centre target after every 10 attempts.

Number of attempts	10	20	30	40	50	60	70	80
Number of centre hits	3	4	10	11	13	16	17	21

a Find the relative frequency of hitting the centre after 10 attempts.

b Find the relative frequency of hitting the centre after 40 attempts.

c Draw a relative frequency polygon for this data.

Hint

Think carefully about what scale you need before you begin to draw your graph. A suitable scale on the vertical would be from 0 to 0.4, and a suitable scale for the horizontal would be from 0 to 80.

d Write down an estimate of the probability that Julia hits the centre target on her next shot.

? **8** A bag contains over 100 counters coloured either red, blue or green.
The ratio of red : blue counters is 2 : 3
The number of green counters is the mean of the number of blue counters and the number of red counters.
Find a possible number of each coloured counter in the bag.

AQA Examination-style questions

AQA Examination-style questions

G

1 A restaurant offers four choices of main course, as shown in the menu.

MENU
Fish and chips
Lasagne
Roast chicken
Beef burger

The following items are ordered between 6pm and 7pm.

Beef burger	Beef burger	Lasagne
Roast chicken	Lasagne	Beef burger
Fish and chips	Lasagne	Beef burger
Roast chicken	Fish and chips	Beef burger
Beef burger	Lasagne	

a Complete a tally chart to represent this information. *(3 marks)*

b Represent this information on a bar chart. *(3 marks)*

F

2 A spinner has ten sections.

Each section is coloured red, blue or green.

It is equally likely to land on red or green.

It is more likely to land on blue than any other colour.

Make a copy of the spinner and label the sections red, blue or green. *(2 marks)*

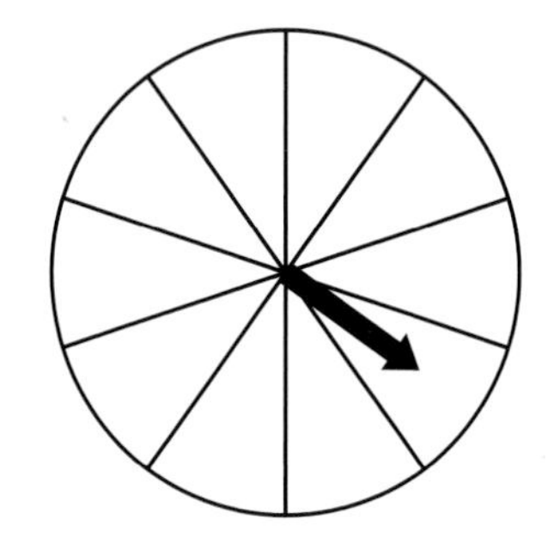

F E

3 Look at the set of five numbers below.

5 6 8 9 10

a Calculate the mean of these numbers. *(2 marks)*

Another number is added to the list.

The range of these six numbers is twice the median of the six numbers.

b Work out the value of the number that has been added.

Show all your working. *(3 marks)*

F

4 The bar chart shows the number of audio books and DVDs a library loaned out last week.

Show that the library loaned out more DVDs than audio books last week.

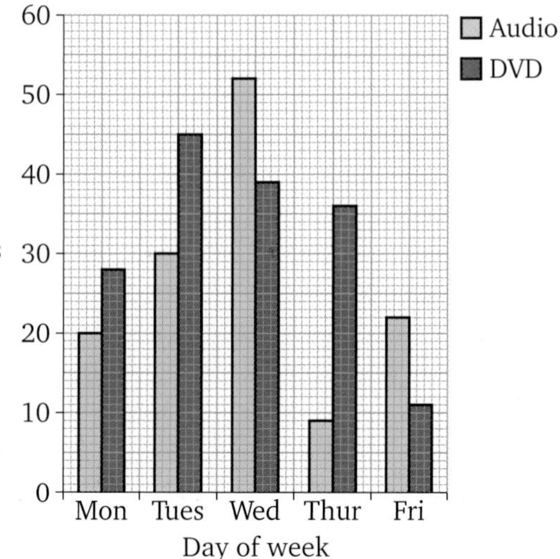

(2 marks)

5 A theatre company is staging a show on four nights.
There are two types of tickets available for the show, full price and concession.
The theatre can seat 300 people.
The number of each type of ticket sold for each show is shown in the table.

	Wednesday	Thursday	Friday	Saturday
Full price	100	150	170	180
Concession	130	50	75	95

a For which day's performance have most tickets been sold? *(1 mark)*
b On which day are more than 50% of the tickets sold concessions? *(1 mark)*
c What percentage of the seats was empty on the Saturday performance? *(3 marks)*
d On a pie chart, represent the number of full-price tickets sold each day.
Label your sections clearly. *(4 marks)*

6 A charity is running a raffle.
The charity sells 250 orange tickets numbered 1 to 250.
The charity sells 170 green tickets numbered 1 to 170.
A ticket is chosen at random to win a holiday.
Find the probability that the ticket selected will be:
a green *(1 mark)*
b numbered 100 *(1 mark)*
c numbered 201. *(1 mark)*

7 A cruise company asks its passengers to complete a short questionnaire.
Here is one of the questions.

Overall how would you rate your holiday?

Design a suitable response section for this question. *(2 marks)*

8 Ann has two fair spinners.

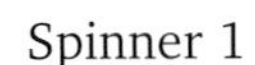

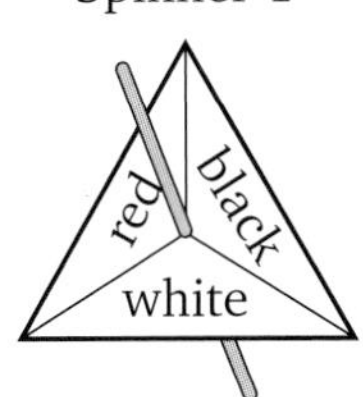

Spinner 2

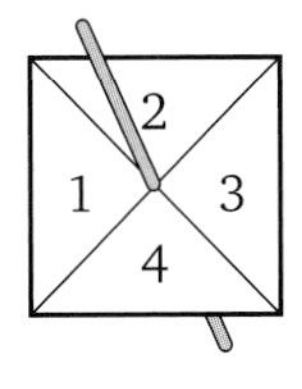

Ann spins both spinners and records the colour and number.
She repeats this a number of times.
a Design a two-way table to show the possible results. *(3 marks)*
b Here are Ann's first five results:
black, 2 red, 2 red, 4 white, 1 black, 2
Put tallies in your table to show Ann's first five results. *(1 mark)*
AQA 2009

D

9 A four-sided spinner is shown.

The spinner is biased.

Some of the probabilities of the spinner landing on a letter are shown in the table.

The probability that the spinner lands on C is double the probability that the spinner lands on D.

The spinner is spun 60 times.

Calculate the number of times you would expect it to land on B.

Letter	Probability
A	0.3
B	
C	
D	0.1

(5 marks)

AQA 2009

10 The table shows the number of bedrooms in each of 1000 houses.

Number of bedrooms	1	2	3	4	5	6
Number of houses	33	124	237	380	186	

The table is incomplete

a How many houses had 6 bedrooms? *(1 mark)*

b State the modal number of bedrooms. *(1 mark)*

c Calculate two other averages for the number of bedrooms per house. *(4 marks)*

D C

11 The table shows the area (in hectares) of 10 towns and their population.

Town	Grimsby	Holmfirth	Lancaster	Longbenton	Macclesfield
Area	2300	900	900	1000	1200
Population	87 000	23 000	46 000	35 000	27 000

Town	Northwich	Otley	Richmond	Rochdale	Widnes
Area	1200	400	200	2200	1600
Population	40 000	15 000	8000	96 000	56 000

a Draw a scatter diagram to represent this information. *(2 marks)*

b Describe the type and strength of the correlation. *(2 marks)*

c Draw a line of best fit on the scatter diagram. *(1 mark)*

d Approximately 29 000 people live in Shipley. Use the line of best fit to estimate the area of Shipley. *(1 mark)*

12 A swimming pool is hosting a charity swim.

People are asked to swim as many lengths as they can in 30 minutes.

The results of 20 people who took part are shown below.

25	33	28	12	15	41	39	32	26	17
8	40	28	36	25	16	22	19	35	27

a Draw an ordered stem-and-leaf diagram to show this information. *(3 marks)*

b Another person swims 24 lengths.

Does this cause the median number of lengths to increase or decrease?

Remember to show your working. *(4 marks)*

C

13 Some year 8 students are taking part in a long-jump event.
The frequency polygon shows the distances jumped by the girls in the event.

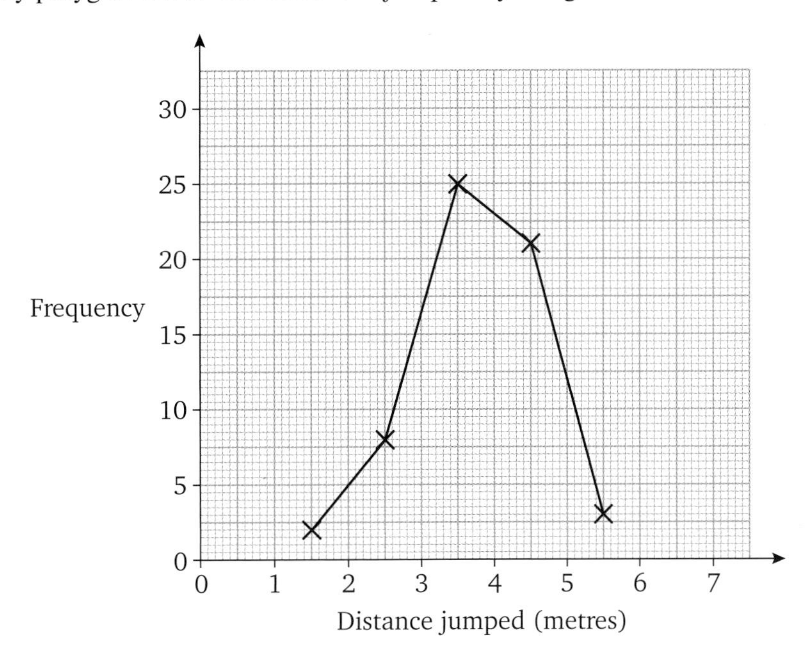

Here is some information about the distances the boys jumped in the same event.

Mean distance	4.80 metres
Shortest jump	2.51 metres
Longest jump	6.95 metres

Hint
Try to make at least one comment about the average and one comment about the range.

Compare the distances jumped by boys and girls.
You should show the result of any calculations you have made. *(4 marks)*

14 A recent newspaper article made the following claim:

Boys receive nearly twice as much pocket money as girls

Mr Chadwick decides to carry out a survey with his class to check the claim.
The results of his survey are shown in the table below.

Amount of pocket money	Number of girls	Number of boys
£0 $< x \leq$ £3	1	0
£3 $< x \leq$ £5	18	7
£5 $< x \leq$ £10	9	22
£10 $< x \leq$ £20	2	1
£20+	0	0

Hint
Think about finding some averages and use them to make comparisons.

Do the results of Mr Chadwick's group support the claim? *(5 marks)*

15 A PE department wishes to compare girls' and boys' fitness levels in Year 7. Write a brief report to explain how the PE department could investigate this.
Your answer should refer to the stages of the data handling cycle. *(4 marks)*

Algebra

1 Sequences and symbols

Unit 2

2 Sequences
5 Working with symbols

Unit 3

3 Working with symbols

Key terms

Write down definitions for the following words. Check your answers in the glossary of your Student Book.

ascending
descending
expand
expression
factorise
like terms
linear sequence
'nth' term
sequence
simplify
substitution
term
term-to-term
unlike terms

Revise... Key points

Sequences (Unit 2)

The terms of a **sequence** follow a rule.

The **term-to-term** rule for the sequence 3, 7, 11, 15, 19, ... is +4 because you add 4 to the last term to get the next one.

This is an **ascending** sequence because the numbers are going up.

The **term-to-term** rule for the sequence 15, 8, 1, −6, −13, ... is −7 because you subtract 7 from the last term to get the next one.

This is a **descending** sequence because the numbers are going down.

Both sequences are called **linear sequences** because the differences are always the same.

The rule to find the next pattern in this sequence is: add a square on the right and add a square on the top.

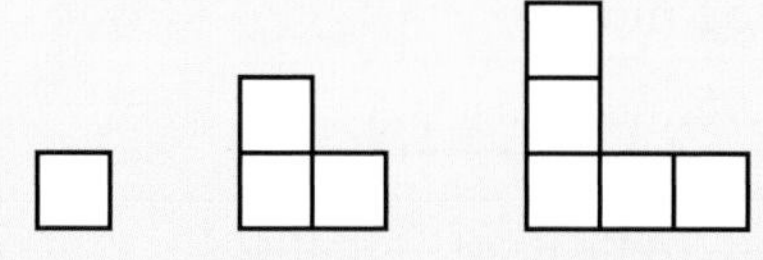

To find the ***n*th term** of a linear sequence, first find the differences.

If the difference is d, the nth term will be $dn + c$, where c is a constant.

Thus, for the sequence 3, 7, 11, 15, 19 ... the nth term will be $4n + c$

To find what is added to $4n$, look at the first term, when $n = 1$.

$4 \times 1 + c = 3$ so $c = -1$ The nth term is $4n - 1$

(Check that substituting $n = 2$ gives you the correct second term.)

Symbols (Units 2, 3)

This is an algebraic **expression**: $3x - 2y - 8z + 5x - z$

It has five **terms**. Each term has its own sign in front.

$(+)3x$ and $+5x$ are **like terms**.

$-2y$ and $-8z$ are **unlike terms**.

To **simplify** an expression, collect like terms: $3x - 2y - 8z + 5x - z = 3x + 5x - 2y - 8z - z$
$= 8x - 2y - 9z$

Algebraic expressions obey the rules of arithmetic:

$2x + 3y = 3y + 2x$ but $2x - 3y \neq 3y - 2x$ $\neq$ is the symbol for 'is not equal to'

$ab = ba$ but $\frac{a}{b} \neq \frac{b}{a}$

You can **substitute** numbers for letters to find the value of an expression.

Example: When $x = -5$ and $y = 4$, $x^2 + y^2 = (-5)^2 + (4)^2 = 25 + 16 = 41$

When you are told to **expand** (or multiply out) brackets, you must multiply all of the terms inside the brackets by the term outside the brackets.

Example: Expand $3a(3 - 2a)$

Solution: $3a(3 - 2a) = 9a - 6a^2$ remember that $3a \times -2a = -6a^2$

Factorising is the opposite of **expanding**.

Example: Factorise $6y^2 + 9y$

Solution: $6y^2 + 9y = 3y(2y + 3)$

Example Sequences 

C

 The first four terms of a sequence are $-8, -3, 2, 7$

Find the nth term.

Bump up your grade

To get a Grade C you need to be able to find the nth term of a sequence.

Solution

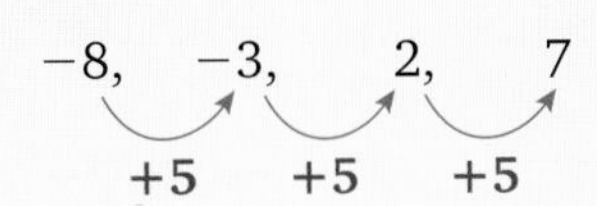

The term-to-term rule is $+5$
so the nth term is $5n + \ldots$

First term: $5 \times 1 + \ldots = -8$ Think: $5 + ? = -8$

Second term: $5 \times 2 + \ldots = -3$ The missing number is -13

Third term: $5 \times 3 + \ldots = 2$ Check this in the second term: $10 + -13 = -3$

The nth term is $5n - 13$

Example Symbols 

E

 **1** When $a = 7$ and $b = -2$, find the value of $5a - 3b^2$

C

 **2** Expand and simplify $4(2x - y) - 3(x - 2y)$

3 This T shape is made of rectangles.

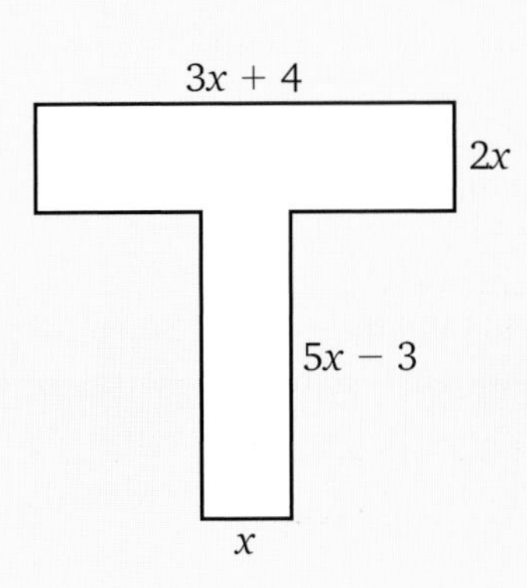

AQA Examiner's tip

Mark the corner where you start adding up the sides for the perimeter. Put a tick beside each side as you write it down – then you will not leave out a side or write down the same side twice.

Write down an expression for the perimeter of the shape.

Solution

1 $5a = 5 \times 7 = 35$ — Working out the first term.

$3b^2 = 3 \times -2 \times -2 = 3 \times 4 = 12$ — Write out b^2 as -2×-2 to get the signs right.

$5a - 3b^2 = 35 - 12 = 23$

AQA Examiner's tip

If you square any number, the answer is always positive.

When you have a negative term (-3) in front of the bracket, remember to change the sign of the second term.

2 $4(2x - y) - 3(x - 2y) = 8x - 4y - 3x + 6y = 5x + 2y$

3 Perimeter $= 3x + 4 + 2x + 3x + 4 + 5x - 3 + 5x - 3 + 2x$
$= 20x + 2$

Practise... Sequences and symbols

Units 2 3

G F

1 The first four terms of a sequence are 2, 8, 14, 20.

a Write down the next two terms of this sequence.

b Write down the term-to-term rule for this sequence.

F

2 $x = 7, y = 2$ and $z = 0$

Find the value of: **a** $2x + 3y - 4z$ **b** xyz **c** $\frac{4x}{y}$

E

3 Simplify: **a** $5a - 5b - a + 2b$ **b** $3y + 9z - 4z - 2y$ **c** $2pq + p + 4qp - 3p$

D

4 The nth term of a sequence is $5n - 1$
Write down the first three terms of the sequence.

5 Expand: **a** $4(2k - 1)$ **b** $m(6 - m)$ **c** $2t(3t + 2)$

6 Factorise: **a** $10x + 5y$ **b** $3k^2 - 9k$

C

7 Write down the nth term for the following linear sequences.

a 2, 5, 8, 11 ... **b** 1, $3\frac{1}{2}$, 6, $8\frac{1}{2}$ ...

8 Expand and simplify $3(2a + 3b) - 5(a - b)$

9 Here is a sequence of numbers: 3, 4, 7, 16.
The rule for continuing this sequence is: multiply by 3 and subtract 5.
What are the next two numbers in this sequence?

10 A sequence has first term 5 and third term 29.
5 ... 29 ... ...
The term-to-term rule for the sequence is to 'multiply by 3 and subtract a'.
Work out the value of a.

11 x and y are integers.
x is odd and y is even.
Decide whether each statement is true or false.

a $x + y$ is always odd.

b $2x - y$ is always odd.

c $5xy$ is always even.

d $2x + 4y$ is always a multiple of 4.

12 $5(x + 6) + 2(x - 8)$ simplifies to $a(x + b)$.
Work out a and b.

Algebra

2 Equations, inequalities and formulae

7 Equations and inequalities
11 Formulae

6 Equations
10 Formulae

Key terms

Write down definitions for the following words. Check your answers in the glossary of your Student Book.

brackets
denominator
equation
expand
expression
formula
inequality
integer
inverse operation
lowest common denominator
operation
solve/solution
subject
substitute
symbol
unknown
value

Revise... Key points

Equations (Units 2 3)

You will be asked to **solve** an equation to find the **unknown**.

For example, the equation $y + 7 = 13$ has the **solution** $y = 6$
Here the unknown is y.

To solve an equation, first think of the **operations** (+, −, ×, ÷) that have been applied to the unknown.

Then reverse the operations, **doing the same to both sides of the equation**.

For example: $5a - 2 = 12$

$5a = 14$ after adding 2 to both sides

$a = 2.8$ after dividing both sides by 5

When the unknown appears on both sides of the equation, collect together on one side all of the terms that contain the unknown. Collect the terms that do **not** contain the unknown on the other side.

For example: $3c + 5 = 1 - c$

$3c + c + 5 = 1 - c + c$ adding c to both sides

$4c + 5 = 1$ after simplifying

$4c + 5 - 5 = 1 - 5$ taking 5 from both sides

$4c = -4$ after simplifying

$c = -1$ after dividing both sides by 4

When the equation contains **brackets**, start by dealing with the brackets. This usually means multiplying out (or **expanding**) the brackets.

For example: $4(d - 3) = 5$

$4d - 12 = 5$ after multiplying out the bracket

$4d = 17$ after adding 12 to both sides

$d = 4.25$ after dividing by 4

When the equation contains fractions, clear the fractions by multiplying by the **lowest common denominator**.

For example: $\frac{2m}{3} - 7 = \frac{m}{6}$

Lowest common denominator is 6 the smallest multiple of both 3 and 6

Multiply every term by 6

$\overset{2}{\cancel{6}} \times \frac{2m}{\cancel{3}} - 6 \times 7 = \cancel{6} \times \frac{m}{\cancel{6}}$ remember to multiply the 7 as well as the fractions

$4m - 42 = m$ after dividing by the common factors

$3m = 42$

$m = 14$

AQA Examiner's tip

Equations with brackets or fractions can be complicated. Make sure each step of your working is clear so you can earn method marks.

Inequalities Unit 2

Learn the four **inequality** symbols.

$<$	$\leqslant$	$>$	$\geqslant$
less than	less than or equal to	greater than	greater than or equal to

Inequalities can be represented on a number line.

This is the number line for $x \geqslant 2$.

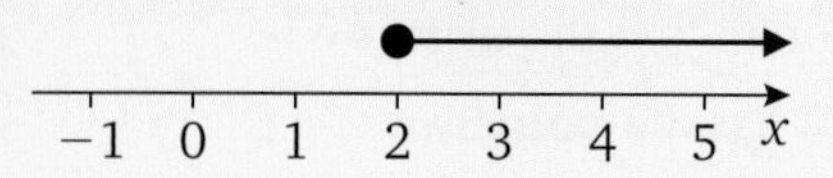

The closed circle shows that the range includes $x = 2$

This is the number line for $x \leqslant 1$ or $x > 3$

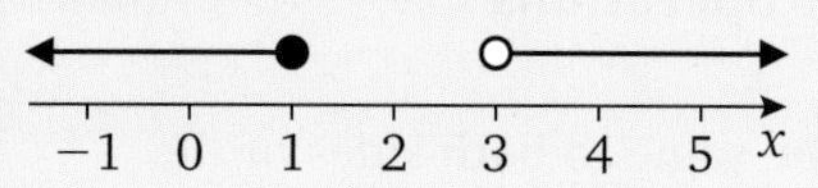

The open circle shows that the range does not include 3.

You may be asked to solve an inequality that looks very similar to an equation.

Follow the same process of reversing the operations.

For example: $4y + 3 < 15$

$4y < 12$ after subtracting 3 from both sides

$y < 3$ after dividing both sides by 4

AQA Examiner's tip

Remember that if you multiply or divide by a negative number you must reverse the inequality.
$4 < 5$ but $-4 > -5$

Hint

Remember an integer is a whole number.

You may be asked to list the **integer** values that satisfy an inequality.

For example: List all the integer values of y such that $-10 < 3y < 6$

Divide the inequality by 3. $-3\frac{1}{3} < y < 2$

−5 −4 −3 −2 −1 0 1 2 3 y

The integer values are $-3, -2, -1, 0, 1$

AQA Examiner's tip

Drawing a number line for $-3\frac{1}{3} < y < 2$ helps you to find all of the values. Remember that zero is an integer.

Formulae Units 2 3

A **formula** may be written in words, such as: Volume of cuboid = length × breadth × height

or in **symbols**: $V = l \times b \times h$

V is on the left-hand side and is called the **subject** of the formula.

This formula tells you how to work out the volume when you know the **values** of other quantities (length, breadth and height). Putting these values into the formula is called **substitution**.

Using the formula above, if $l = 10$, $b = 5$ and $h = 3$ then $V = 10 \times 5 \times 3 = 150$

You may be asked to change the subject of a formula, for example if you want to find the length of a cuboid and you know the breadth and height. The formula $V = l \times b \times h$ has to be changed round so that l is the subject.

To change the subject of a formula, use the same steps as used in solving an equation.

$V = l \times b \times h$ divide both sides by b and h to leave l on its own

$\frac{V}{bh} = l$ now swap the sides to get l on the left

$l = \frac{V}{bh}$

Distinguishing formulae, expressions and equations 

Like an **equation**, a formula always includes an equals sign. A formula is true for a range of values, whereas equations are true only for certain values which you find by solving the equation.

An **expression** does not contain an equals sign.

$5x + 2$ is an expression.

$5x + 2 = 17$ is an equation: it is only true when $x = 3$

$P = 5x + 2$ is a formula for finding P when you know the value of x.

Example Equations Unit 2

E

Jodie thinks of a number, doubles it and adds 11.

Her answer is 25.

Call Jodie's number x.

Write down an equation in x and solve it to find Jodie's number.

Solution

Double x means $2x$.

$2x + 11 = 25$

$2x = 14$ after taking 11 from both sides

$x = 7$ after dividing both sides by 2

Jodie's number is 7.

Check the answer by working through with the number 7.

Double it to get 14. Add 11 to get 25. ✓

Example Equations 

C

Triangle PQR is isosceles. One angle is $x°$. Another angle is $7x°$.

Find the two possible values of x.

Solution

In an isosceles triangle two angles are equal.

The angles could be $x°$, $x°$ and $7x°$ or they could be $7x°$, $7x°$ and $x°$.

$x + x + 7x = 180$	$7x + 7x + x = 180$
$9x = 180$	$15x = 180$
$x = 20$	$x = 12$

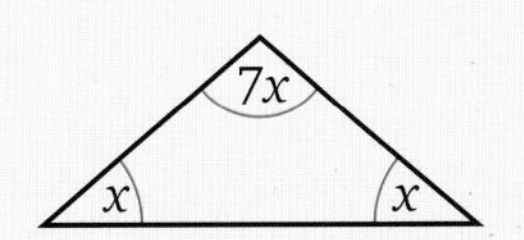

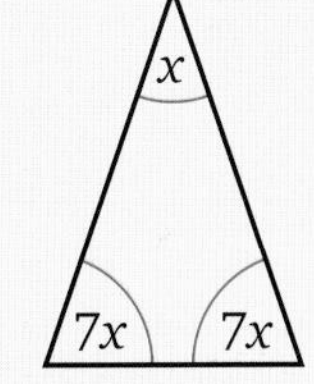

Check the first answer: $20 + 20 + 140 = 180$ ✓

Check the second answer: $84 + 84 + 12 = 180$ ✓

AQA Examiner's tip

When the question involves geometry, it is always helpful to draw a diagram.

Example Equations Unit 2

C

Solve the equation $3(2x - 5) = 13 - 2(5 - x)$

Solution

Start by multiplying out the brackets.

$6x - 15 = 13 - 10 + 2x$ Remember that -2 in front of the second bracket multiplies $-x$ to become $+2x$

$6x - 15 = 3 + 2x$

Collect terms with x on the left and terms without x on the right.

$6x - 2x = 3 + 15$

$4x = 18$

$x = 4\frac{1}{2}$

Bump up your grade

To get a Grade C you have to be able to solve equations with brackets.

Example Inequalities Unit 2

D

Write down the inequalities shown by these number lines.

a

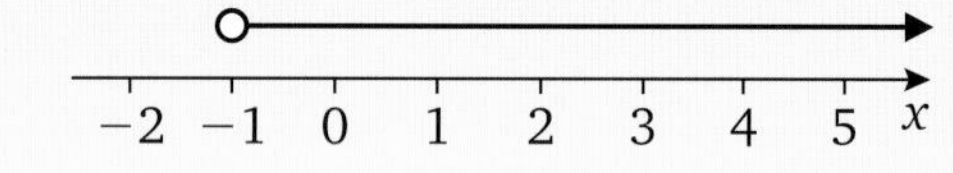

b

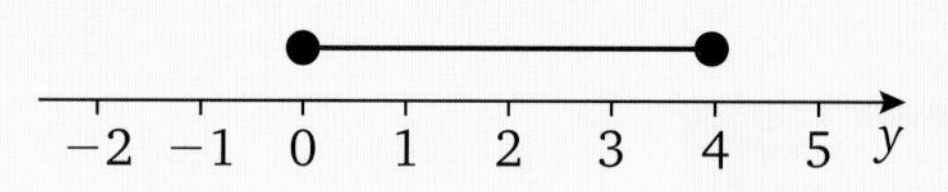

Solution

a There is an open circle at $x = -1$ so that point is not included.
The arrow goes from -1 to greater values.
This is the number line for $x > -1$

b There are closed circles at $y = 0$ and at $y = 4$ so these points are included.
The line goes from 0 to 4.
This is the number line for $0 \leqslant y \leqslant 4$

Example Writing a formula Units 2 3

F

Matt is 5 years younger than Dean.
Write this:

a as a formula in words

b as a formula in symbols.

Solution

a Matt's age = Dean's age minus 5 this tells us how to work out Matt's age if we know Dean's age

b Use m for Matt's age and d for Dean's age. m is the subject of the formula
$m = d - 5$

Example Using a formula in words 

G F

SupaCarHire charges £25 per day plus £45 returnable deposit.

a James books a car for three days.
How much does he pay?

b Peggy pays £270 when she books a car.
How many days' hire has she booked?

Solution

a James pays 3 × £25 + £45 = £120 make sure you add 45 **after** you have worked out 3 × 25

b £270 − £45 = £225 the £45 deposit has to be taken away from £270 first
Number of days × 25 = 225 this requires a division
225 ÷ 25 = 9
Peggy has booked the car for 9 days.

Example Substitution in a formula

Units 2 3

E

The total surface area of the cuboid is given by
$S = 2xy + 2yz + 2xz$

Find the total surface area when $x = 12$, $y = 5$ and $z = 4$

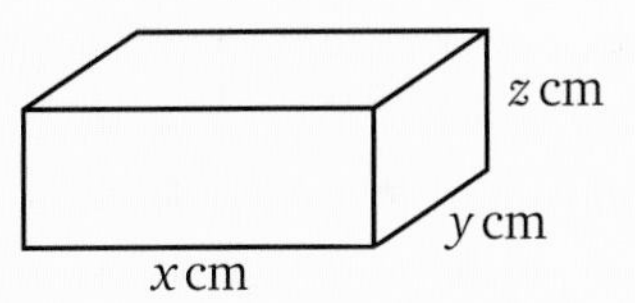

Solution

$2xy = 2 \times 12 \times 5 = 120$

$2yz = 2 \times 5 \times 4 = 40$

$2xz = 2 \times 12 \times 4 = 96$

$S = 120 + 40 + 96 = 256$

The total surface area is $256\,\text{cm}^2$.

Hint

It is a good idea to work out one term at a time and then add them up.

AQA Examiner's tip

Make sure you earn method marks by setting out your working clearly.

Example Changing the subject of a formula

Unit 2

C

Make B the subject of the formula $P = 2L + 2B$

Solution

$P = 2L + 2B$ write the formula down first

$P - 2L = 2L + 2B - 2L$ subtract $2L$ from both sides

$P - 2L = 2B$

$\frac{P - 2L}{2} = B$ divide both sides by 2

$B = \frac{P - 2L}{2}$ swap sides to get B on the left

Bump up your grade

To get a Grade C you need to be able to change the subject of a formula.

Practise... Equations, inequalities and formulae

G F

1 **a** Ramin travels 3 miles by taxi.
How much does he pay?

b Freddie's taxi journey costs him £16.
How many miles has he travelled?

Value Taxis

We charge you £4 basic fare plus £1.50 per mile

F

2 Jill's journey to college takes x minutes.

a Minnie's journey takes 5 minutes longer.
Write down an expression in x for the length of Minnie's journey.

b Nina's journey takes twice as long as Jill's journey.
Write down an expression in x for the length of Nina's journey.

3 Solve these equations:

a $5a = 30$ **b** $b + 3 = 2$ **c** $c - 4 = 0$ **d** $4d = 18$

F E

4 **a** Write down a formula for the perimeter, p, of a square of side y.

b Write down a formula for the perimeter, P, of this octagon.

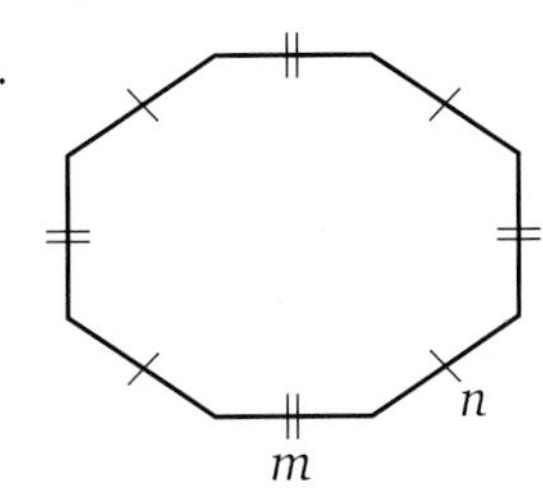

E

5 John's weekly wage is a basic £195 plus £8 commission for every TV he sells.
Write down an expression for his total wage, £T, in a week when he sells n TVs.

6 Solve these equations:

a $\frac{e}{3} = 6$ **b** $2f - 6 = 8$ **c** $4 + 2g = 3$ **d** $15 = 3 - 4h$

7 Kim thinks of a number, multiplies it by 5 and subtracts 7.
The answer is 33.
Write down an equation and solve it to find Kim's number.

D

8 Solve these equations:

a $3k - 1 = 4 + 2k$ **b** $m + 13 = 7 - 2m$ **c** $6n - 5 = 2n - 3$

9 The angles of a quadrilateral are $3x°$, $5x°$, $(x + 70)°$ and $(x - 20)°$.
Write down an equation and solve it to find the value of x.

10 Old Mother Hubbard buys x bags of apples and y bags of pears.
Write down an expression for the total cost.

Small Apples
large bag for only £1.50

Best Pears
£2 a bag

11 Show each of these inequalities on a number line:

a $x > 0$ **b** $y \leqslant 1$ **c** $-5 \leqslant n < -1$

F E C

12 A formula for the sum of the interior angles of a polygon is

Double the number of sides, subtract four, then multiply by 90°

a Use this formula to find the sum of the interior angles of a polygon with 5 sides.

b Write the formula in numbers and symbols, using k for the number of sides and starting S = …

c The sum of the interior angles of a k-sided polygon is 1080°.
Work out the value of k.

C

13 Solve these inequalities:

a $5x + 7 \geqslant 2$ **b** $4 - 3y < 10$

14 List all the integer values of n such that $-11 < 5n \leqslant 15$

15 Solve these equations:

a $9x + 2 = 4(2x - 1)$ **b** $3(2y - 1) = 4 - y$ **c** $3(t - 1) - 5(t + 2) = 7$

16 Make p the subject of the formula $t = 3p - 7$

17 The formula for the perimeter, P, of a rectangle is $P = 2l + 2w$ where l is the length and w is the width.

a Make w the subject of the formula.

b Work out the value of w when $P = 29$ and $l = 8$

18 Ryan thinks of a number, doubles it, adds 15, then multiplies the result by 3.
The answer is 69.
Write down an equation and solve it to find Ryan's number.

19 Solve these equations:

a $4 + \frac{x}{3} = 2$ **b** $\frac{2y + 5}{6} = 3$ **c** $\frac{5z}{8} - \frac{7z}{12} = 2$

Algebra

3 Trial and improvement

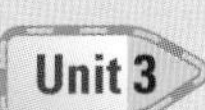

13 Trial and improvement

Key terms

Write down definitions for the following words. Check your answers in the glossary of your Student Book.

round

solve/solution

trial and improvement

Trial and improvement

Trial and improvement is used to **solve** equations by using **estimations** which get closer and closer to the **solution**.
On the examination paper you will be told when to use trial and improvement. Do **not** use it, for example, to solve linear equations (such as $5x - 3 = 2 + x$). You will be asked to give a **rounded** answer – read the question carefully to make sure your answer has the correct degree of accuracy.

It is best to set out your working in a table, as shown in the example.

Example Trial and improvement

C

Use trial and improvement to find the solution of the equation $x^3 - 5x = 62$ that lies between 4 and 5.

Give your answer to one decimal place.

Bump up your grade

To get a Grade C you have to be able to use trial and improvement accurately to find the solution of a cubic equation.

Solution

Set up the table, with the three headings shown below.

Start by trialling the values given in the question, $x = 4$ and $x = 5$.

x	$x^3 - 5x$	Comment
4	$64 - 20 = 44$	too small
5	$125 - 25 = 100$	too large

These results suggest that the solution is closer to 4 than to 5, because 44 is closer than 100 to 62.

For your next trial, you could use $x = 4.3$ or 4.4

Some students prefer to always go halfway between, at $x = 4.5$

Here are two more trials.

4.3	$79.507 - 21.5 = 58.007$	too small
4.4	$85.184 - 22 = 63.184$	too large

AQA *Examiner's tip*

Write out your calculations clearly, with at least two decimal places.

Marks will be given for this working, so you must make sure the examiner can read it.

The number 62 lies between 58.007 and 63.184

So the solution lies between 4.3 and 4.4

To find out whether it is closer to 4.3 or 4.4, trial at $x = 4.35$

halfway between 4.3 and 4.4

4.35	$82.312\ldots - 21.75 = 60.562\ldots$	too small

The number 62 lies between 60.562… and 63.184

So the solution lies between 4.35 and 4.4

To one decimal place, the solution is 4.4

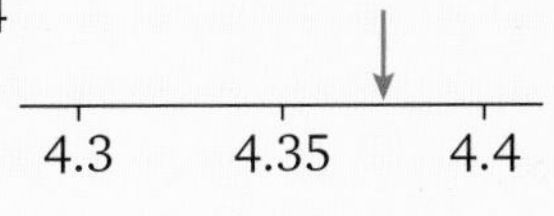

Practise... Trial and improvement Unit 3

C

1 A solution of the equation $x^3 + x = 8$ lies between $x = 1$ and $x = 2$
Use trial and improvement to find this solution to one decimal place.

2 A solution of the equation $y^3 - 3y = 7$ lies between $y = 2$ and $y = 3$
Use trial and improvement to find this solution to one decimal place.

3 A solution of the equation $t^3 - t^2 = 52$ lies between $t = 4$ and $t = 5$
Use trial and improvement to find this solution to two decimal places.

4 A solution of the equation $m^3 - 7m = 5$ lies between $m = -3$ and $m = -2$

a Use trial and improvement to find this solution to one decimal place.

b Use trial and improvement to find a positive solution of this equation.
Give your answer correct to one decimal place.

5 A solution of the equation $x^2 + \frac{2}{x} = 19$ lies between $x = -5$ and $x = -4$
Use trial and improvement to find this solution to one decimal place.

6 A solution of the equation $p^3 - 3p^2 = 175$ lies between $p = 6$ and $p = 7$
Use trial and improvement to find this solution to one decimal place.

Hint

Remember that $3p^2 = 3 \times p \times p$,
not $3p \times 3p$.

7 The dimensions of this cuboid are in centimetres.
The volume of the cuboid is 120 cm^3.
Find the value of y, correct to one decimal place.

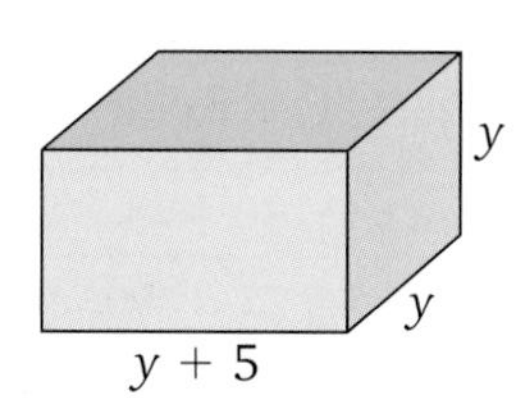

Algebra

4 Coordinates and graphs

Key terms

Write down definitions for the following words. Check your answers in the glossary of your Student Book.

axis (pl. axes)
coefficient
coordinates
gradient
horizontal axis
linear
line segment
midpoint
origin
quadrant
speed
variable
vertical axis

Revise... Key points

Coordinates Unit 2

This diagram shows a pair of **axes**, the **origin**, O, and the point P, with the **coordinates** $(-3, 4)$.

The **horizontal axis** is the x-axis and the **vertical axis** is the y-axis.

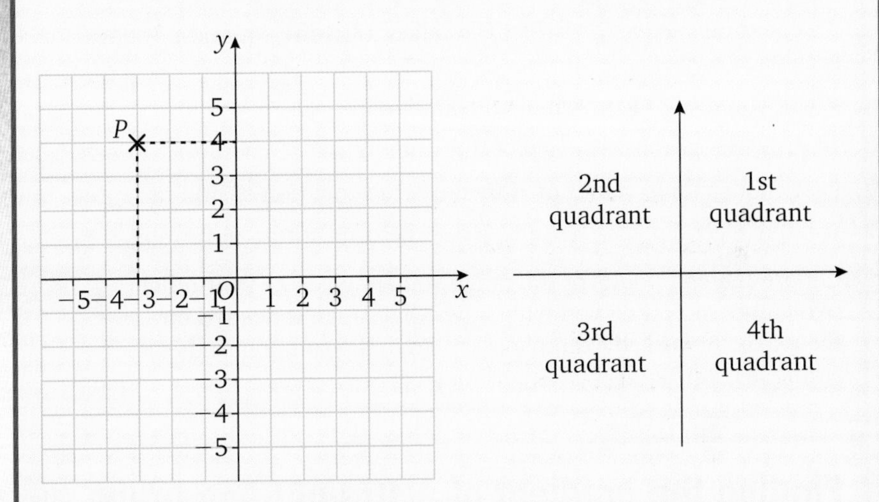

The axes divide the grid into four **quadrants** as shown in the second diagram.

The y-axis is the line $x = 0$ because all the points on it have 0 as their x-coordinate.

Lines parallel to the y-axis have equations like $x = 1$, or $x = -5$ because all points on them have the same x-coordinate.

The x-axis is the line $y = 0$ because all the points on it have 0 as their y-coordinate.

Lines parallel to the x-axis have equations like $y = 3$, or $y = -4$ because all points on them have the same y-coordinate.

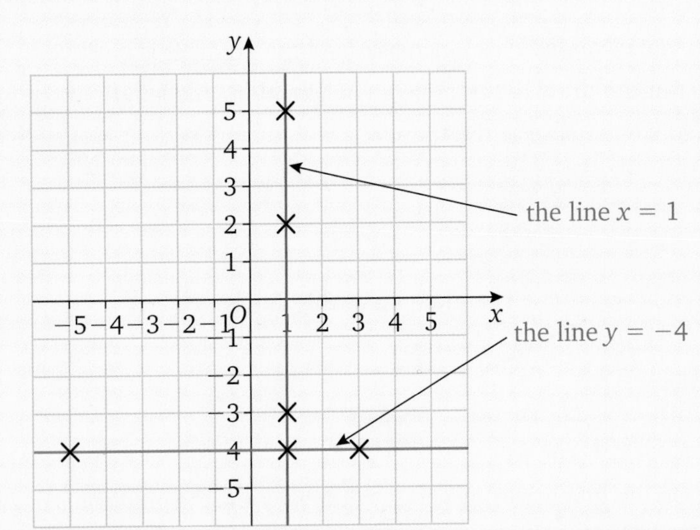

The points $(1, 2)$, $(1, 5)$, $(1, -3)$ all lie on the line $x = 1$

The points $(1, -4)$, $(3, -4)$, $(-5, -4)$ all lie on the line $y = -4$

To find the **midpoint** of a **line segment**, find the mean of the coordinates of the endpoints.

So the midpoint of the line joining

(3, 7) to (5, −1) is $\left(\frac{3+5}{2}, \frac{7+-1}{2}\right) = (4, 3)$

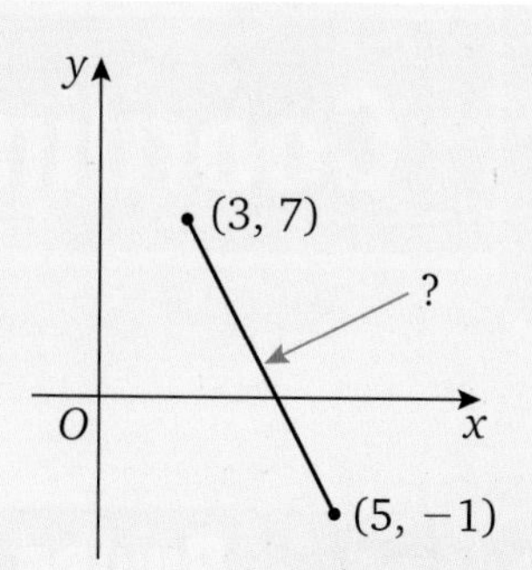

AQA Examiner's tip

It is always a good idea to sketch a diagram showing the position of the two endpoints.

This helps you to see whether your answer is sensible.

Bump up your grade

To get a Grade C you need to be able to work out the coordinates of a midpoint.

Drawing straight-line graphs Unit 2

An equation such as $y = 4 - 5x$ is called a **linear** equation. It does not contain any powers of x or y, such as x^2 or y^3.

When you draw the graph of an equation like this, it will always be a straight line.

To draw the graph, always plot three points whose coordinates fit the equation of the line.

First choose three values of x, then use the equation to find the values of y.

$x = 0$ is a good choice because it is easy to substitute in the equation.

Choose the end points of the range you are given as your other two values of x, then you will know what range is needed on the y-axis. A worked example is shown in the solution to the Example on straight-line graphs on page 84.

AQA Examiner's tip

If your three plots are **not** in a straight line, you have made a mistake! Go back and check your working for finding the value of y.

Finding the gradient of a straight-line graph Unit 2

The **gradient** of a straight-line graph is a measure of how steep it is.

You can find the gradient from the graph.

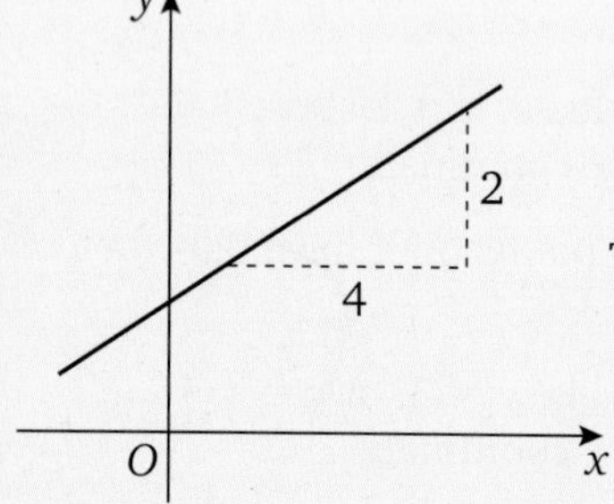

The gradient of this line is $\frac{2}{4} = \frac{1}{2}$

AQA Examiner's tip

When you are working from the graph, check the scales used on the graph as they may not be the same on both axes.

As x increases y also increases and the gradient is positive.

You can draw the triangle anywhere along the length of the graph.

Make the base of the triangle a whole number so it is easy to do the division.

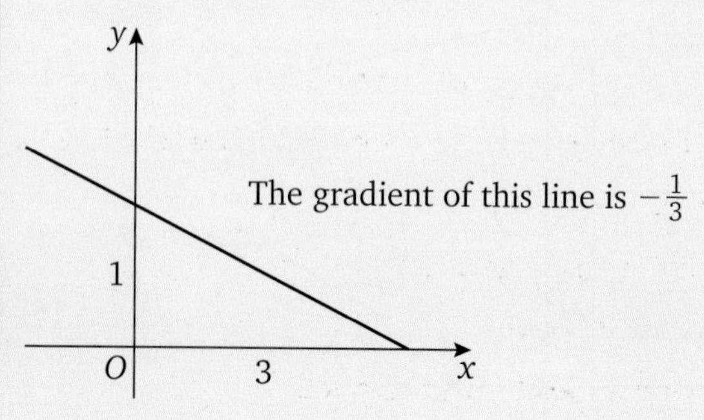

The gradient of this line is $-\frac{1}{3}$

Bump up your grade

To get a Grade C you need to be able to work out the gradient of a line.

On this graph, as x increases y decreases and the gradient is negative. The line slopes from top left to bottom right.

Conversion graphs Units 2 3

You can use a conversion graph to convert from one unit of measurement to another, for example dollars to pounds or inches to centimetres.

Plot three points to draw the graph.

You can then read other values from the graph.

This is a graph for converting inches to centimetres.

To draw a graph like this you use the fact that 12 inches is approximately 30 centimetres (and that 0 inches = 0 centimetres).

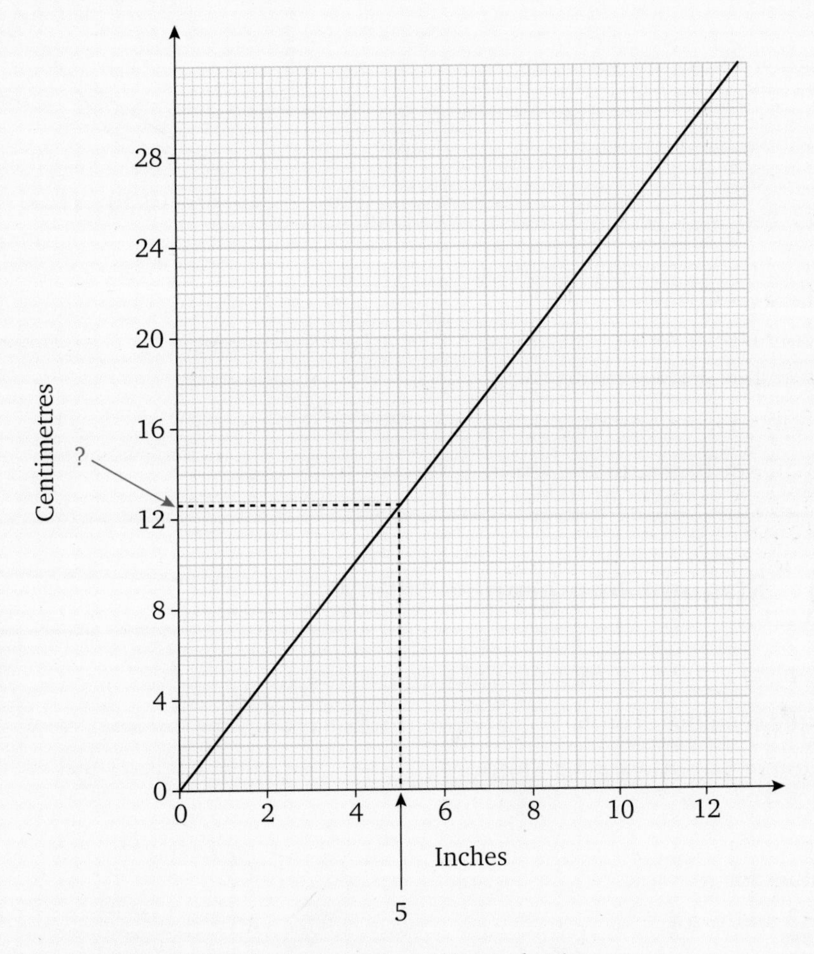

To convert 5 inches to centimetres, start from 5 on the horizontal axis.

Draw a vertical line up to meet the graph.

Draw a horizontal line from this point across to the vertical axis.

Read off the result, which is 12.6 cm.

Distance–time graphs Units 2 3

Distance–time graphs tell you about a journey.

Time is measured along the horizontal axis and distance is measured along the vertical axis.

Time may be measured from the start or be given as actual time using am and pm or the 24-hour clock.

The distance is measured from a particular point, usually the starting point.

If the graph goes back to the horizontal axis, the traveller is returning to the starting point.

The gradient of the line on a distance–time graph gives the **speed** of travel.

If the graph is horizontal (no increase in the distance direction) the traveller has stopped.

Example Distance–time graphs

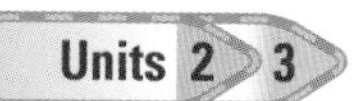
Units 2 3

E D

The journeys of three people are shown on the graph.

a Who is travelling the fastest? How can you tell from looking at the graph?

b How far apart are Ben and Chris after 2 hours?

c At what time does Alice overtake Ben?

d Work out Alice's speed in km/h.

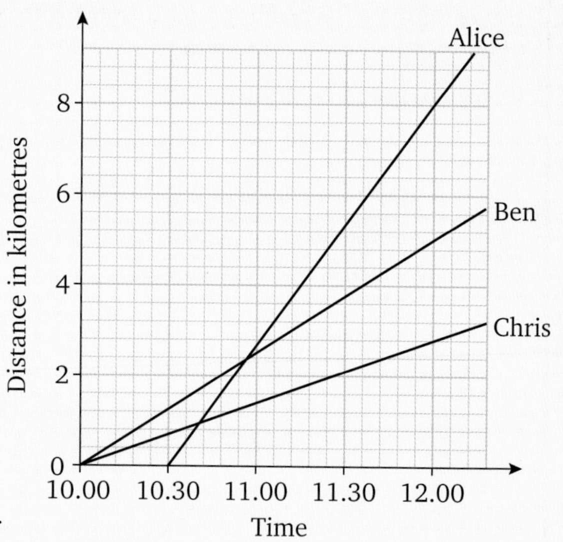

Solution

a The steepest line shows the fastest speed, so Alice is travelling the fastest.

b On the vertical scale, one small square $= \frac{1}{5}$ of 2 km

$= 0.4$ km.

After 2 hours, Ben and Chris are $6 \times 0.4 = 2.4$ km apart.

c To find when Alice overtakes Ben, read off the time at the point where their graphs cross.

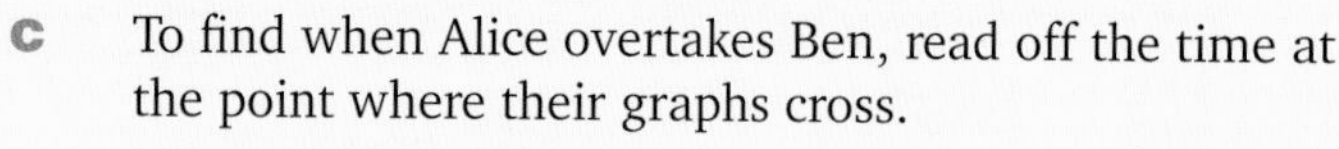
On the horizontal scale, one small square $= \frac{1}{5}$ of 30 mins $= 6$ mins

Alice overtakes Ben at about 10.56.

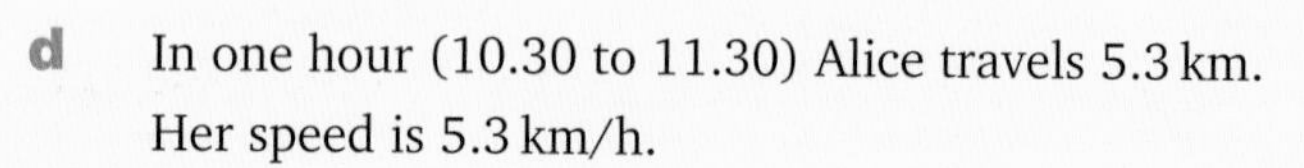
d In one hour (10.30 to 11.30) Alice travels 5.3 km.

Her speed is 5.3 km/h.

Example Straight-line graphs

Unit 2

D C

a Draw the graph of $y = 2 - 3x$ for values of x from -3 to 4.

b Write down the coordinates of the point where this line crosses the line $y = -6$

c Use your graph to find the gradient of the line.

Solution

a Choose 3 values for x:

$x = 0$, $x = -3$ and $x = 4$

x	-3	0	4
y	11	2	-10

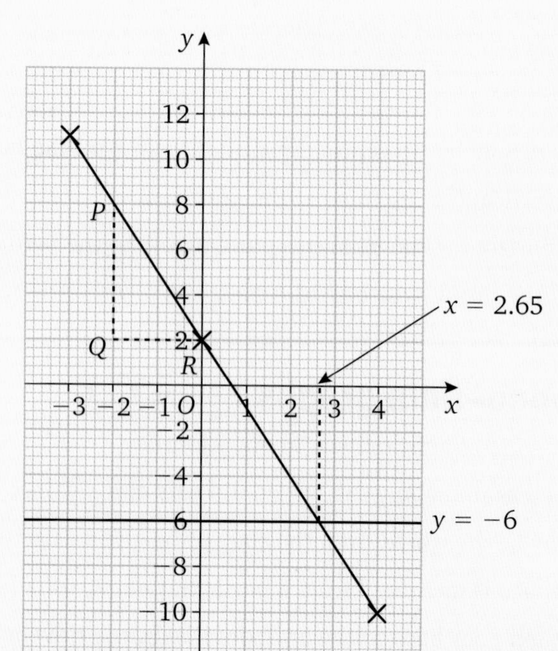

b The graph crosses $y = -6$ at $(2.65, -6)$.

c Use triangle PQR to find the gradient.

$PQ = 6$ units, $QR = 2$ units

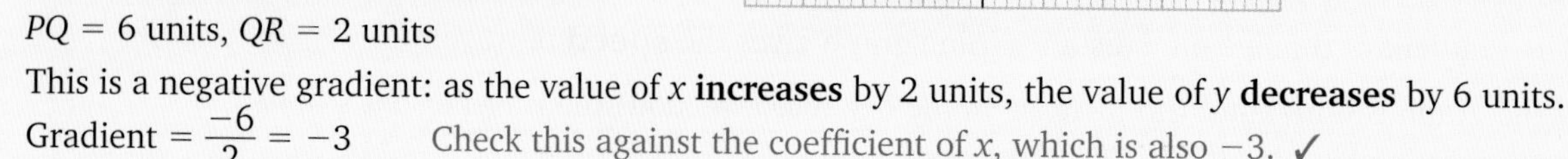
This is a negative gradient: as the value of x **increases** by 2 units, the value of y **decreases** by 6 units.

Gradient $= \frac{-6}{2} = -3$ Check this against the coefficient of x, which is also -3. ✓

Example Distance–time graph Unit 2

This graph shows Ginny's journey to work.
She leaves home at 07:00 and cycles to the station.
She catches a train to the town where she works.
When she gets off the train she walks to her office.

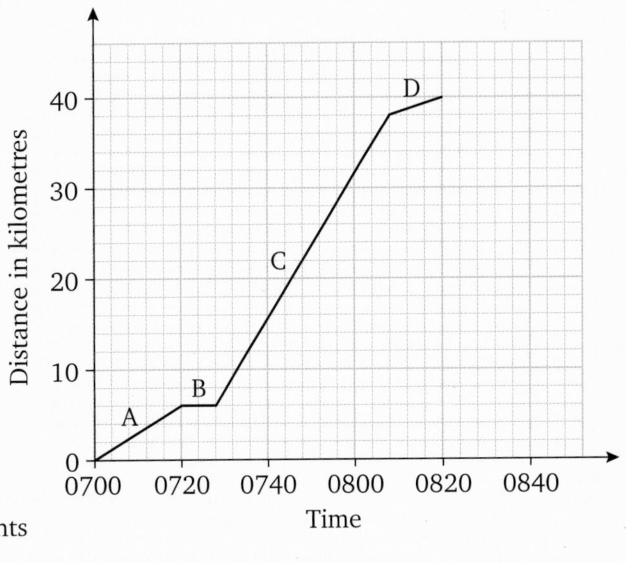

a How far is it from Ginny's home to the station?

b How long does she wait for the train?

c How far does she travel on the train?

d Work out the average speed of the train in km/h.

Solution

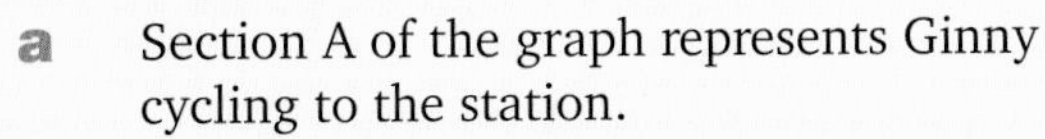

a Section A of the graph represents Ginny cycling to the station.
On the vertical scale each small square represents 2 km, so the distance is 6 km.

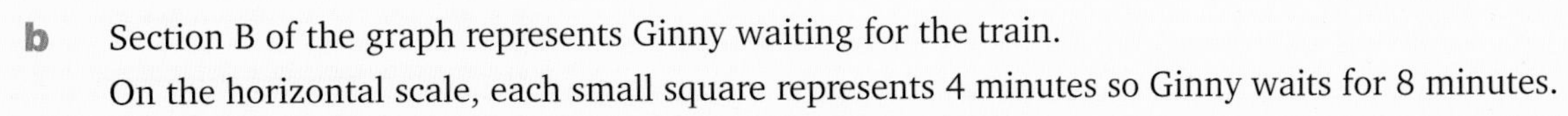

b Section B of the graph represents Ginny waiting for the train.
On the horizontal scale, each small square represents 4 minutes so Ginny waits for 8 minutes.

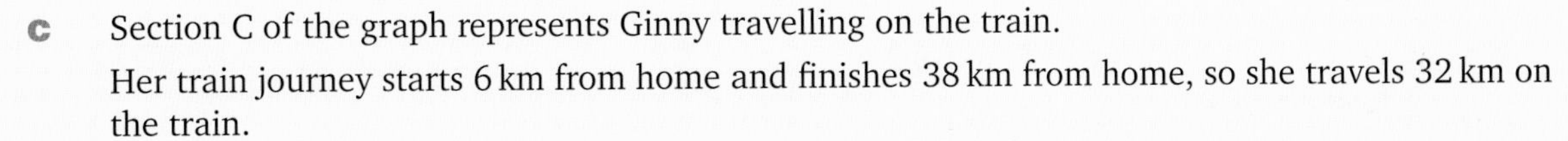

c Section C of the graph represents Ginny travelling on the train.
Her train journey starts 6 km from home and finishes 38 km from home, so she travels 32 km on the train.

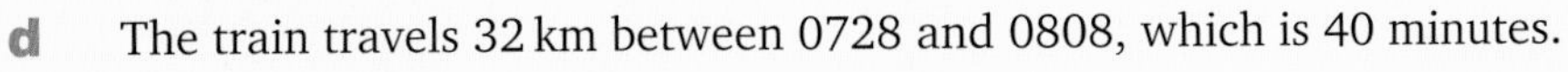

d The train travels 32 km between 0728 and 0808, which is 40 minutes.

$$\text{Speed} = \frac{\text{Distance}}{\text{Time}} = \frac{32}{\frac{40}{60}}\text{ km/h} = 48\text{ km/h}$$

AQA Examiner's tip

It is often easier to work out speeds using proportions.
In 40 minutes the train travels 32 km.
In 20 minutes it travels 16 km.
In one hour it travels $3 \times 16 = 48$ km.
So the speed is 48 km/h.

Bump up your grade

To get a Grade C you need to be able to work out average speed from a distance–time graph using distance travelled in a number of minutes.

Practise... Coordinates and graphs

Units 2 3

G F E D C

1 Write down the coordinates of each of the points A to F.

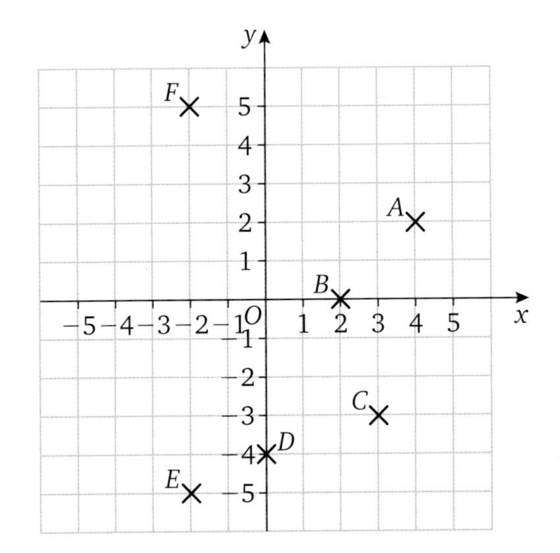

F E

2 This graph shows the cost of electrical repairs, based on a call-out fee plus an hourly rate.

a What is the call-out fee?

b What is the cost of a repair lasting $1\frac{1}{2}$ hours?

c Liam paid £180. How long did his repair take?

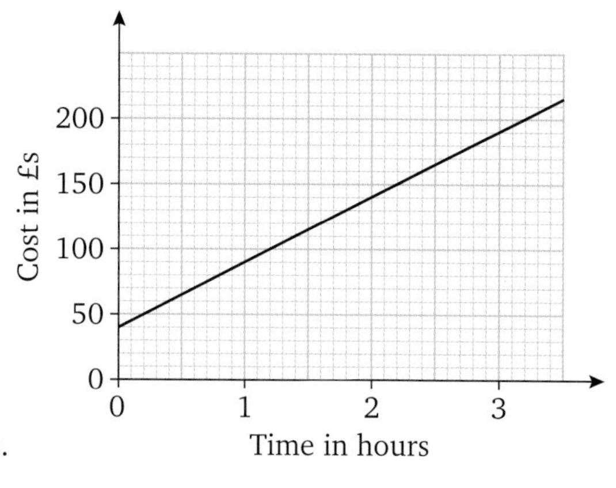

3 Draw a set of axes labelled from −6 to 6 on both axes.
Plot the points $A(0, 3)$, $B(4, -1)$ and $C(-2, -1)$.
These points are three vertices of a parallelogram.
Find the three possible positions for the fourth vertex.
Write down the coordinates of these three positions.

E

4 Write down the equation of the line that goes through the points (1, 5) and (−2, 5).

5 **a** Complete this table of values for $4x + 3y = 24$

b Draw the graph of $4x + 3y = 24$ for values of x from 0 to 6.

x	0	3	
y			0

6 Greg leaves home at 1100 to walk to Aqaville, 8 km away.
The line on the graph represents his journey.

a How far from home is he at 1200?

b Describe what happens at 1148.

c What is Greg's average speed before 1148?

Greg's brother, Harry, leaves their home at 1200 and cycles to Aqaville at 18 km/h.

d Copy the graph and draw a line to represent Harry's journey.

e When and where does Harry overtake Greg?

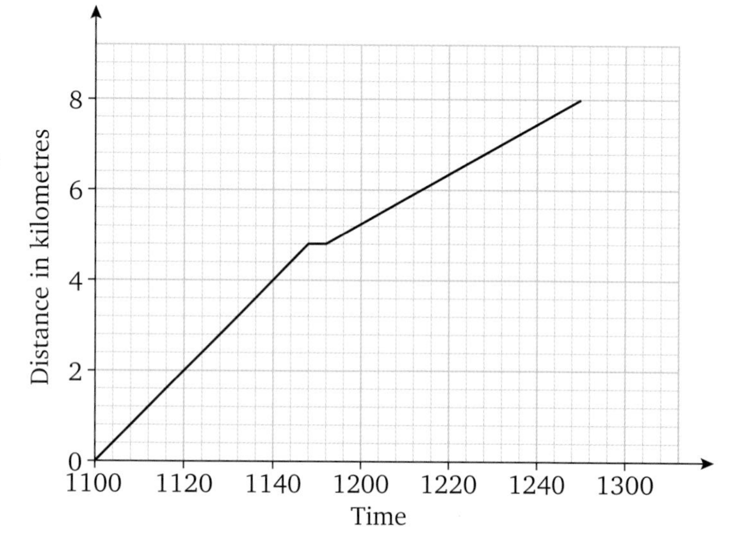

E D C

C

7 **a** Work out the coordinates of the midpoint of the line PQ where P is (−3, 2) and Q is (5, 3).

b $M\ (1\frac{1}{2}, -2)$ is the midpoint of the line AB.
The coordinates of A are (−1, −4).
Find the coordinates of B.

8 Work out the gradient of the line shown on this grid.

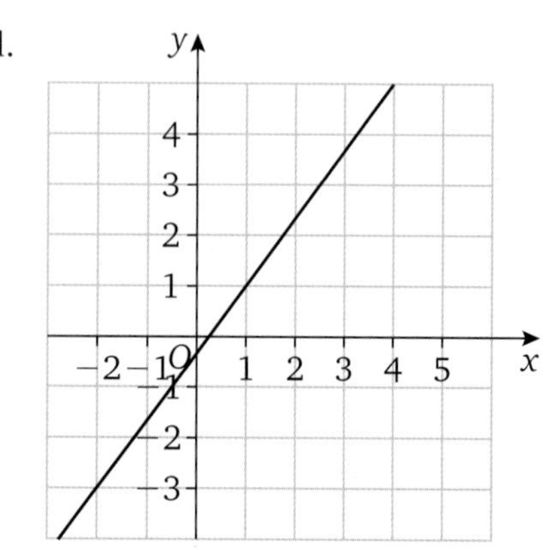

Algebra

5 Quadratic functions

Unit 3
17 Quadratics

Key terms

Write down definitions for the following words. Check your answers in the glossary of your Student Book.

line of symmetry
parabola
quadratic equation
quadratic expression
symmetrical

Revise... Key points

A **quadratic expression** has x^2 as its highest power.

For example: $3x^2$, $x^2 - 5$, $x^2 + 4x$, $x^2 + x - 1$, $6 - x^2$, $7 - 3x + 2x^2$

The graph of a quadratic expression is a curve, called a **parabola**.

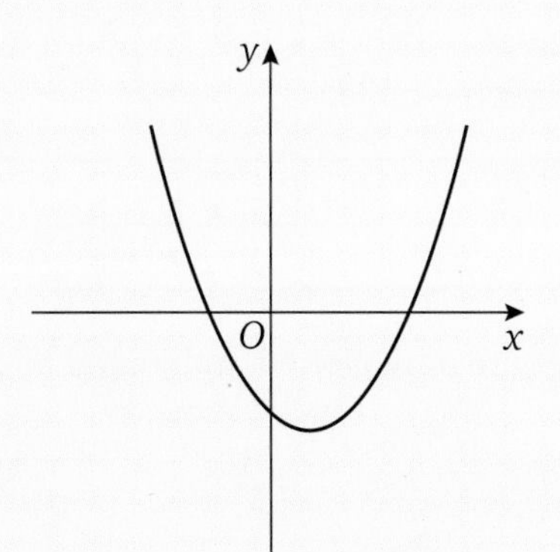

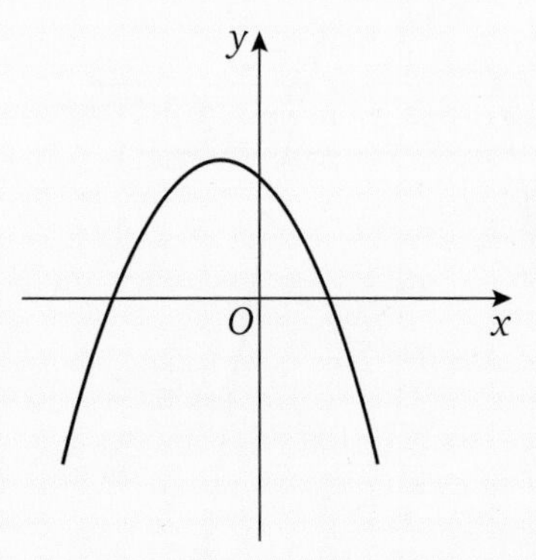

If the coefficient of x^2 is positive, the parabola will look like the letter U.

If the coefficient of x^2 is negative, the parabola will look like a hill (or an upside-down U).

All parabolas have a **line of symmetry**.

To draw the graph of a quadratic, you will need to plot many points and join them with a **smooth curve**.

The coordinates of your points will come from a table of values.

In the exam, you will usually be given a table with some values already worked out.

Example Drawing graphs of quadratic functions Unit 3

D

a Copy and complete the table of values for $y = 2 - x^2$

x	−3	−2	−1	0	1	2	3
y	−7		1			−2	

b Draw the graph of $y = 2 - x^2$ for values of x from −3 to 3.

c Use your graph to find the values of x when $y = -1$

Solution

a When $x = -2, y = 2 - (-2)^2 = 2 - 4 = -2$

When $x = 0, y = 2 - (0)^2 = 2 - 0 = 2$

When $x = 1, y = 2 - (1)^2 = 2 - 1 = 1$

When $x = 3, y = 2 - (3)^2 = 2 - 9 = -7$

Here is the completed table.

x	−3	−2	−1	0	1	2	3
y	−7	−2	1	2	1	−2	−7

AQA Examiner's tip

It is likely that the bottom line of your table will show symmetry – look out for this to help you check your calculations.

b The coordinates are now plotted and joined by a smooth curve.

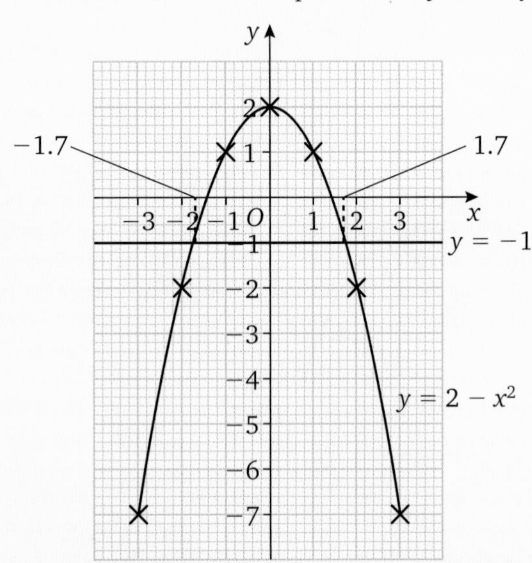

Draw a smooth curve through the top three points – never draw a pointed shape at the top or bottom of a parabola.

Check that your curve is the shape you expect.

The coefficient of x^2 is –1 so the curve is an upside-down U.

It is **symmetrical** about the y-axis.

You can see this symmetry in the bottom row of the table.

c Draw the line $y = -1$, which crosses the curve at two points.

Read off the values of x at these points by drawing lines up to the x-axis.

The values of x are −1.7 and 1.7.

AQA Examiner's tip

You will lose marks if you fail to draw a smooth curve that goes through every plotted point.

DO NOT:

- join the plots with straight lines
- draw multiple lines between the points
- draw wobbly lines
- draw a pointed top or bottom to the graph.

If your plots do not join up to make a U or a hill, look to see which plot seems out of place and re-do your working for that value of x.

Bump up your grade

To get a Grade C you need to be able to read values from your graph.

You will be given marks for answers that are accurate to within one small square of your graph paper.

Example Drawing graphs of quadratic functions Unit 3

C

a Copy and complete the table of values for $y = x^2 - 3x - 2$

x	−2	−1	0	1	2	3	4
y	8		−2	−4		−2	

b Draw the graph of $y = x^2 - 3x - 2$ for values of x from −2 to 4.

c Write down the coordinates of the points where the curve crosses the x-axis.

d Draw the line of symmetry of the parabola and write down its equation.

Solution

a When $x = -1, y = (-1)^2 - (3 \times -1) - 2 = 1 + 3 - 2 = 2$

When $x = 2, y = (2)^2 - (3 \times 2) - 2 = 4 - 6 - 2 = -4$

When $x = 4, y = (4)^2 - (3 \times 4) - 2 = 16 - 12 - 2 = 2$

Here is the completed table.

x	−2	−1	0	1	2	3	4
y	8	2	−2	−4	−4	−2	2

b The coordinates are now plotted and joined by a smooth curve.

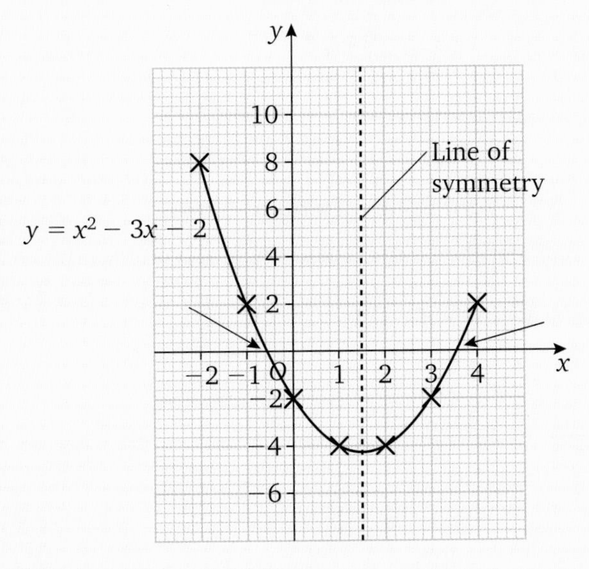

The coefficient of x^2 is 1, so we expect a U shape.
This curve is not a complete U, but we can see that if the x-axis was extended, the rest of the U would appear.

AQA Examiner's tip

You must never produce a parabola with a flat bottom (or top).
In this example, you have to draw a curve that goes below (1, −4) and comes back to (2, −4). It should not look pointed and you should try to make it look symmetrical.
Remember: Flat bottoms lose marks.

c The curve crosses the x-axis at (−0.6, 0) and (3.6, 0).
(Shown by the arrows on the diagram.)

d The line of symmetry goes through the lowest point (the minimum) on the curve.
The equation of the line of symmetry is $x = 1.5$

Practise... Quadratic functions Unit 3

D C

D

1 **a** Copy and complete the table of values for $y = x^2 + 5$

x	−3	−2	−1	0	1	2	3
y	14		6	5		9	

b Draw the graph of $y = x^2 + 5$ for values of x from −3 to 3.

c Use your graph to find the values of x when $y = 8$

2 **a** Copy and complete the table of values for $y = 3x - x^2$

x	−2	−1	0	1	2	3	4
y	−10		0	2			−4

b Draw the graph of $y = 3x - x^2$ for values of x from −2 to 4.

c Write down the coordinates of the maximum point on the graph.

3 **a** Copy and complete the table of values for $y = 3x^2 - 7$

x	−3	−2	−1	0	1	2	3
y	20	5			−4	5	

Hint

Be careful when you work out values of $3x^2$.
Square x, then multiply by 3.
When $x = -2$, $3x^2 - 7 = 3 \times (-2)^2 - 7 = 3 \times 4 - 7 = 12 - 7 = 5$

b Draw the graph of $y = 3x^2 - 7$ for values of x from −3 to 3.

c Write down the coordinates of the points where the graph crosses the x-axis.

C

4 This is Pete's table of values for $y = x^2 - 5x + 1$

x	−2	−1	0	1	2	3	4	5	6
y	15	7	1	3	−5	−5	−3	1	7

a Plot the values shown in the table.
One of the plots has been calculated incorrectly.

b Correct this plot and draw the graph.

c Write down the coordinates of the minimum point on the graph.

5 **a** Copy and complete the table of values for $y = (x + 1)(x - 5)$

x	−2	−1	0	1	2	3	4	5
y	7		−5	−8		−8	−5	

b Draw the graph of $y = (x + 1)(x - 5)$ for values of x from −2 to 5.

c Use your graph to find the values of x when $y = -3$.

6 **a** Copy and complete the table of values for $y = 1 + 2x - x^2$

x	−1	0	1	2	3	4
y		1	2		−2	−7

b Draw the graph of $y = 1 + 2x - x^2$ for values of x from −1 to 4.

c Write down the equation of the line of symmetry of this parabola.

AQA Examination-style questions

1 **a** Write down the next two terms of this sequence.

5 9 13 17 21 … … *(2 marks)*

b Write down the rule for continuing this sequence. *(1 mark)*

c What is the 10th number in the sequence? *(1 mark)*

2 **a** Write down the next two terms of this sequence.

20 17 14 11 8 … … *(2 marks)*

b Write down the rule for continuing this sequence. *(1 mark)*

c Ramin says the sequence stops after these two extra terms.
Explain why Ramin is wrong. *(1 mark)*

3 An approximate rule for changing from Celsius (C) to Fahrenheit (F) is:

double C and add on 30

a Find the value of F when C = 12 *(2 marks)*

b Find the value of C when F = 84 *(2 marks)*

4 Jack offers a washing machine repair service.
He works out his charges using this formula: £35 for the call-out plus £32 per hour.

a Jack is called out to repair Mrs Cook's washing machine.
It takes him an hour and a half.
How much does Jack charge Mrs Cook? *(2 marks)*

b Jack charges Mr Stone £123.
How long did it take him to repair Mr Stone's washing machine?
Give your answer in hours and minutes. *(3 marks)*

5 Point P is shown on the diagram. Copy the diagram.

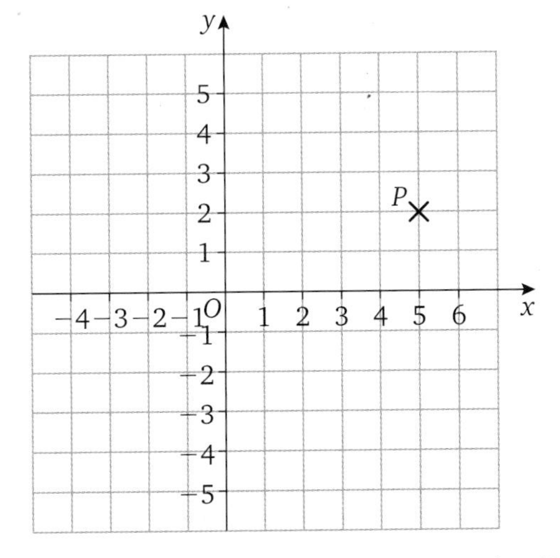

a On your diagram, plot the point $Q(2, 4)$ and the point $R(-2, -2)$. *(2 marks)*

b $PQRS$ is a rectangle.
Mark the point S and write down its coordinates. *(2 marks)*

6 Simplify:

a $4m - 2m + m$ *(1 mark)*

b $8x + 5y - 3x - y$ *(2 marks)*

c $3k^2 - 5k + 1 - k + 6$ *(2 marks)*

7 Solve the equations:

a $5x = 15$ *(1 mark)*

b $y + 7 = 3$ *(1 mark)*

c $\frac{z}{3} = 6$ *(1 mark)*

d $4t - 3 = 11$ *(2 marks)*

8 This is a conversion graph for metres and feet.

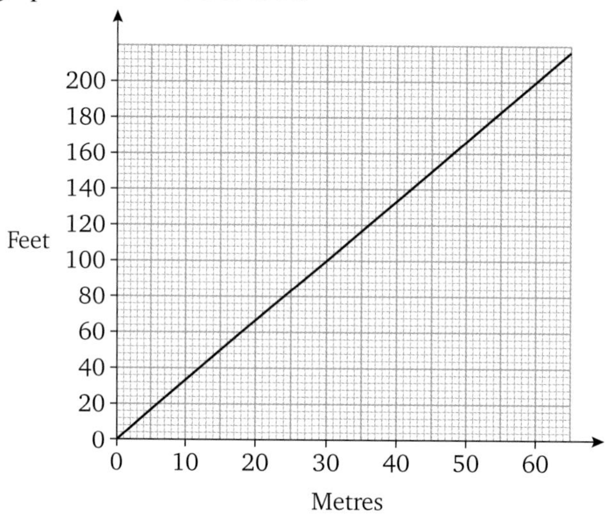

a Use the graph to convert 60 feet to metres. *(1 mark)*

b Use the graph to convert 45 metres to feet. *(1 mark)*

c Convert 180 metres to feet. *(2 marks)*

9 Write down an expression for the perimeter of this shape.

Give your answer in its simplest form. *(2 marks)*

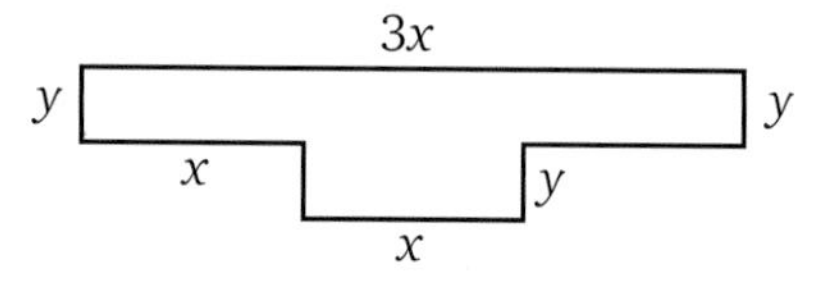

10 a Copy and complete the table of values for $y = 10 - 2x$ *(1 mark)*

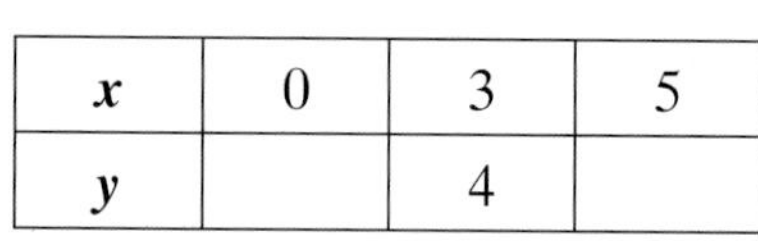

x	0	3	5
y		4	

b Copy the grid and draw the graph of $y = 10 - 2x$ for values of x from 0 to 5. *(2 marks)*

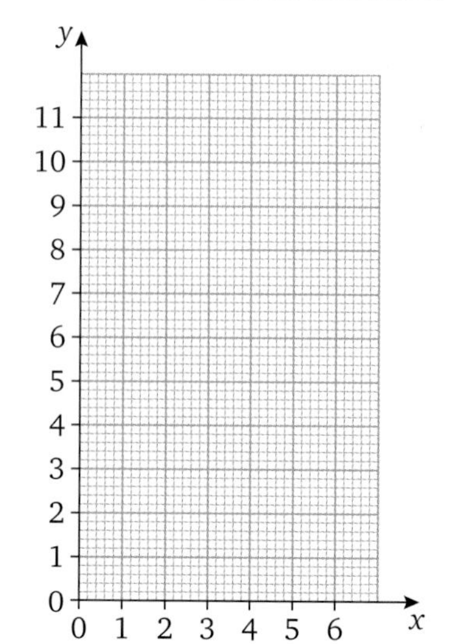

c On the same grid, draw the graph of $y = 2x$ *(2 marks)*

d Write down the coordinates of the point where the two graphs cross. *(1 mark)*

11 Jodie uses matchsticks to make patterns of squares.

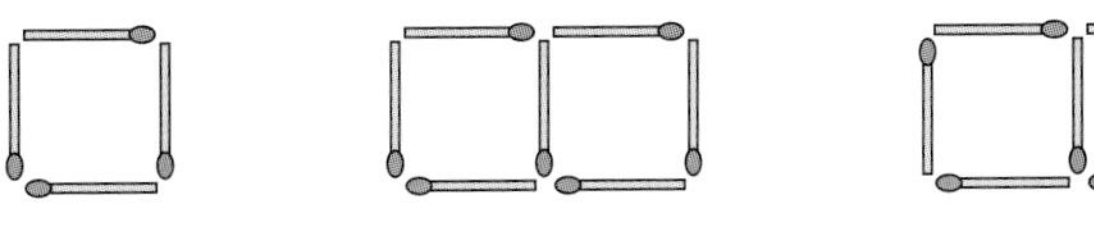

Pattern 1 Pattern 2 Pattern 3

a Draw Pattern 4. *(1 mark)*

b How many matchsticks are in Pattern 4? *(1 mark)*

c There are 73 matchsticks in Pattern 24.
How many matchsticks are in Pattern 25? *(1 mark)*

d Work out the number of squares in the pattern that uses 37 matchsticks. *(2 marks)*

e How many matchsticks will Jodie use to make a pattern of n squares? *(2 marks)*

12 Kate thinks of three numbers.

Her first number is x.

Her second number is five larger than her first number.

Her third number is twice her first number.

a Write down her second and third numbers in terms of x. *(2 marks)*

The sum of her three numbers is 61.

b Write down an equation in x. *(1 mark)*

c Solve the equation to find the value of x. *(2 marks)*

13 Shaun cycles from Ayton to Beeville. The graph shows his journey.

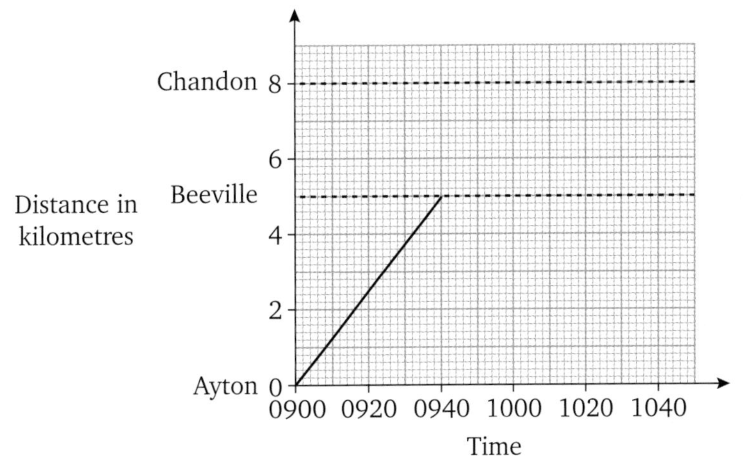

a How far is it from Ayton to Beeville? *(1 mark)*

Shaun stays at Beeville for 20 minutes.
Then he cycles on to Chandon at the same average speed as before.

b Copy the grid and show the rest of his journey on it. *(3 marks)*

c What time does he arrive at Chandon? *(1 mark)*

d Work out Shaun's average speed between Ayton and Beeville. *(3 marks)*

14

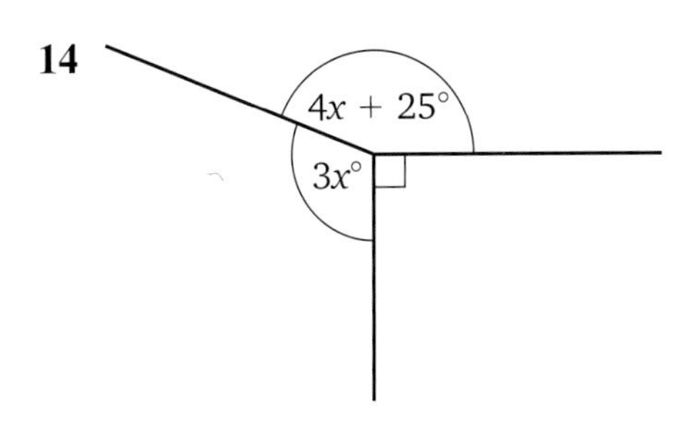

Form an equation and solve it to find the value of x. *(3 marks)*

D

15 a Copy and complete the table of values for $y = x^2 - 5$

x	-3	-2	-1	0	1	2	3
y	4		-4	-5		-1	

(2 marks)

b Copy the grid. Draw the graph of $y = x^2 - 5$ for values of x from -3 to $+3$.

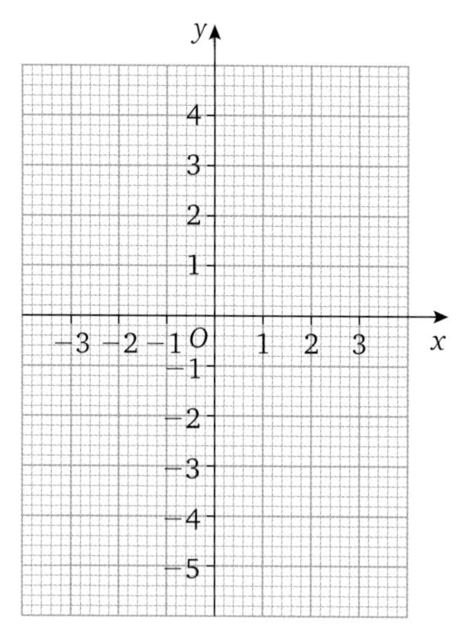

(2 marks)

c Write down the values of x at the points where the line $y = -2$ crosses your graph. *(2 marks)*

16 Which of the words **expression**, **equation**, **formula** describes each of the following?

a $V = \pi r^2 h$ *(1 mark)*

b $5x - 2$ *(1 mark)*

c $T = 45w + 20$ *(1 mark)*

d $4y + 3 = 7 - 2y$ *(1 mark)*

e $8 = \frac{2p - 1}{3}$ *(1 mark)*

f $p + 2q + 3r - 5t$ *(1 mark)*

17 Factorise:

a $7x - 21$ *(1 mark)*

b $2y - y^2$ *(1 mark)*

D
C

18 Solve the equations:

a $2p + 11 = 5 - p$ *(3 marks)*

b $5(3 - q) = 20$ *(3 marks)*

c $4t - 7 = 3(t + 5)$ *(3 marks)*

C

19 The cooking time, T minutes, for a joint of beef is given by the formula $T = 30W + 15$, where W is the weight in kilograms. Make W the subject of the formula. *(2 marks)*

20 a List all the integer solutions of the inequality $-3 < 2n \leqslant 6$ *(3 marks)*

b Solve the inequality $5y - 3 > 9$ *(2 marks)*

21 A solution of the equation $x^3 + 2x = 39$ lies between $x = 3$ and $x = 4$.

Use trial and improvement to find this solution, correct to one decimal place.

You must show all your trials. *(3 marks)*

Geometry and measure

1 Angles

Unit 3

2 Angles
7 Properties of polygons

Key terms

Write down definitions for the following words. Check your answers in the glossary of your Student Book.

acute angle
alternate angles
bearing
bisect
corresponding angles
diagonal
decagon
equilateral triangle
exterior angle
hexagon
interior angle
isosceles triangle
nonagon
obtuse angle
octagon
parallel
pentagon
perpendicular
polygon
quadrilateral
reflex angle
right angle
right-angled triangle
scalene triangle
straight angle
triangle
vertically opposite angles

Revise... Key points

Naming angles Unit 3

Angles are measured in degrees. The sign for degrees is °.

A full turn is 360°

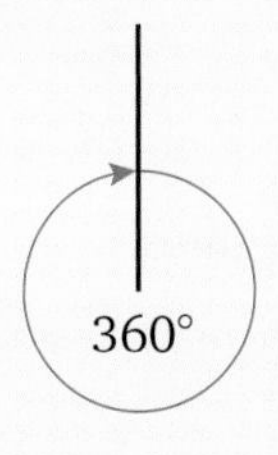

A straight line is a half turn or 180°

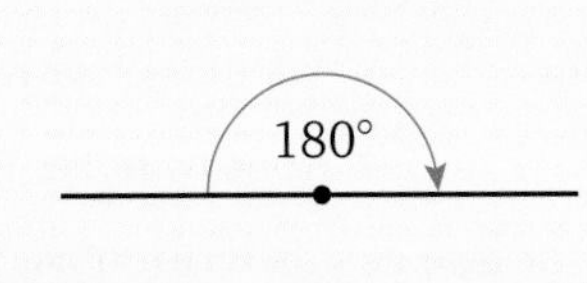

A quarter turn, called a **right angle**, is 90°

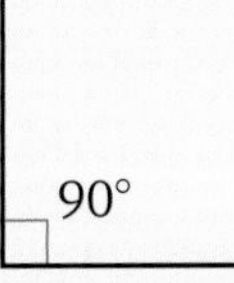

Acute angles are less than 90°

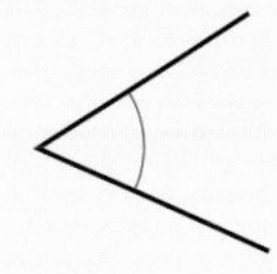

A right angle is 90°, and is marked with a square on the corner of the angle

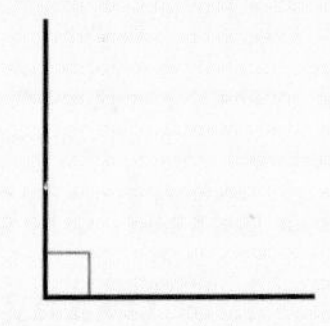

Obtuse angles are between 90° and 180°

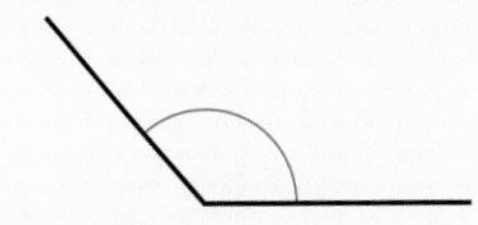

A **reflex angle** is greater than 180°

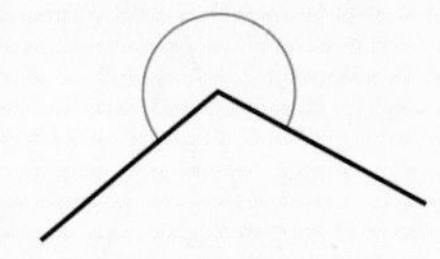

Calculating angles Unit 3

Angles at a point add up to 360°

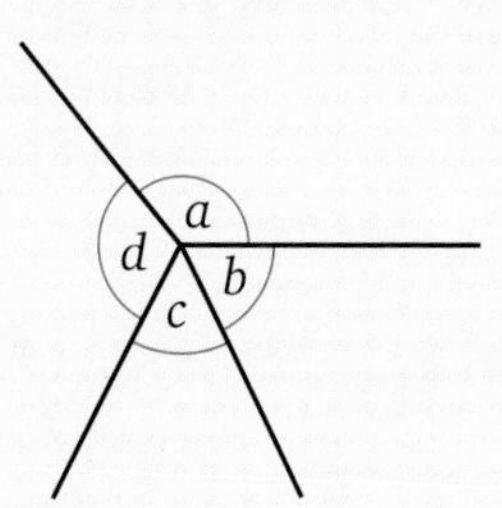

$a + b + c + d = 360°$

Angles on a straight line add up to 180°.

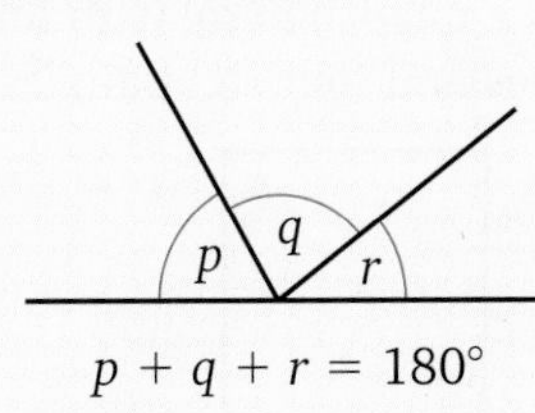

$p + q + r = 180°$

Where two lines cross, the **vertically opposite angles** are equal.

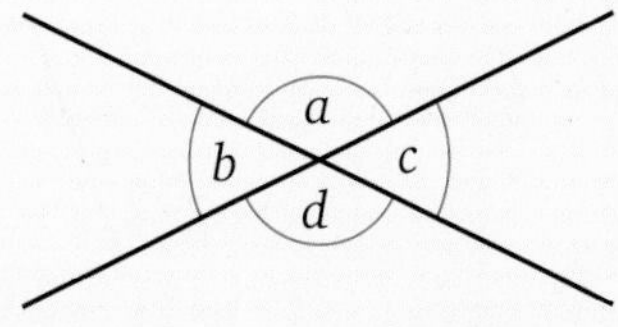

$a = d$ and $b = c$ (vertically opposite angles)

Parallel lines are marked with arrows.

Alternate angles are equal

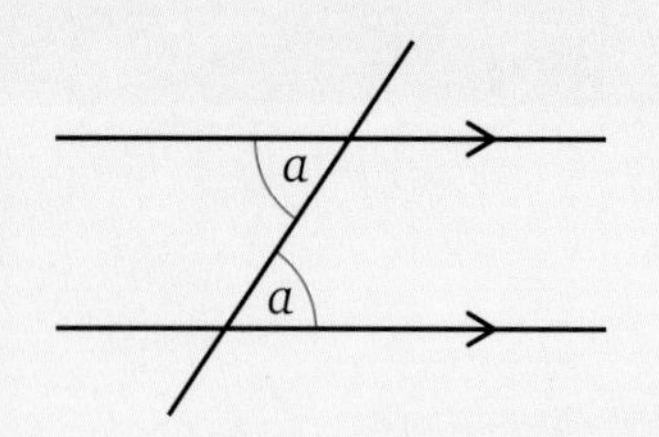

Corresponding angles are equal

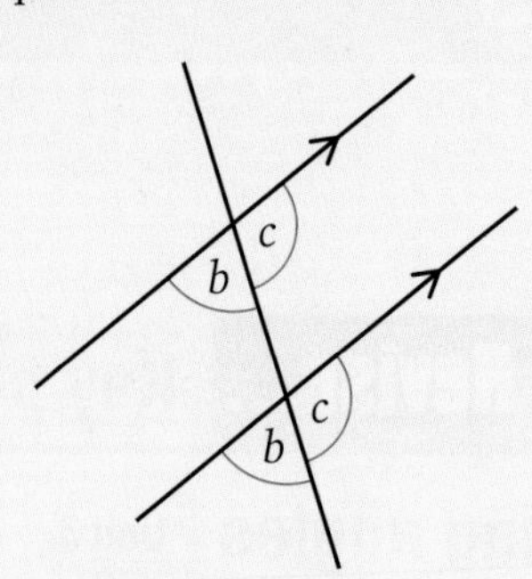

Interior angles (or allied angles) add up to 180°

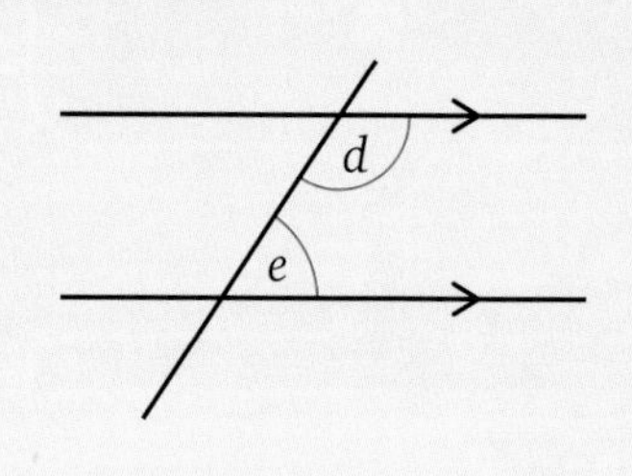

Bearings Unit 3

Directions can be described using three-figure **bearings**.

A three-figure bearing is the angle measured clockwise from north.

Angles less than 100° need a zero in front to make three figures.

For example, the bearing for due east is written as 090° instead of 90°.

The bearing of Y from X is a. The bearing of X from Y is b.

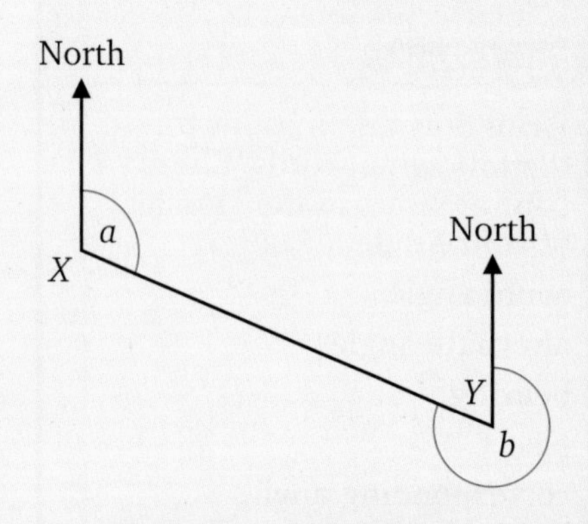

Angles and shapes Unit 3

The angles in a **triangle** add up to 180°.

An **isosceles triangle** has two equal sides and two equal angles.

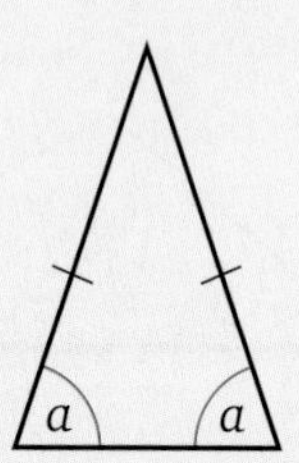

An **equilateral triangle** has three equal sides and three equal angles of 60°.

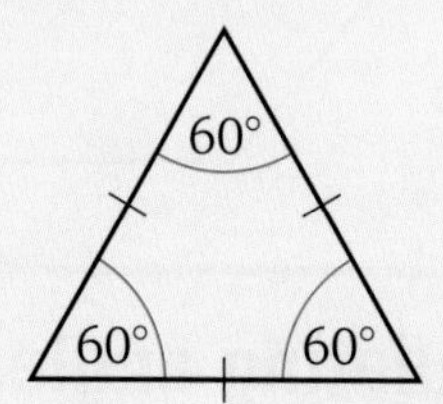

In a **scalene triangle** all three sides are different lengths and all three angles are different sizes.

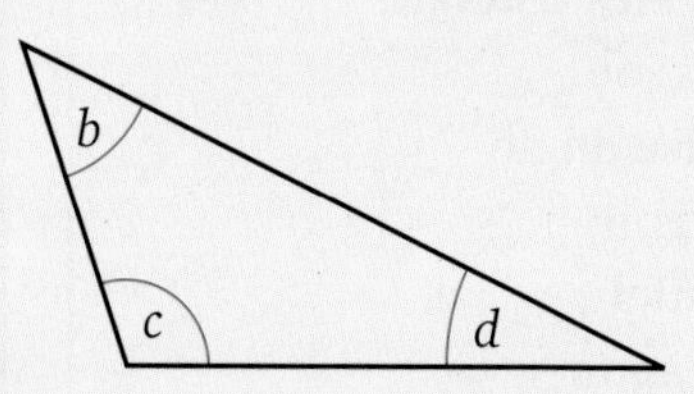

Quadrilaterals

Square – a quadrilateral with four equal sides and four right angles

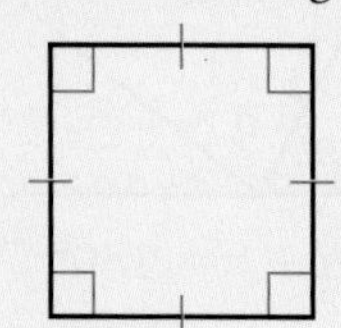

Trapezium – a quadrilateral with one pair of parallel sides

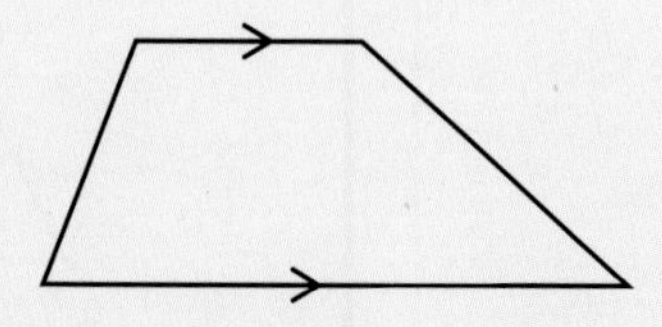

Rectangle – a quadrilateral with four right angles, and opposite sides equal in length

Isosceles trapezium – a trapezium where the non-parallel sides are equal in length

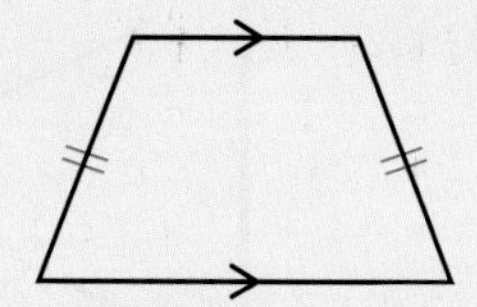

Kite – a quadrilateral with two pairs of equal adjacent sides

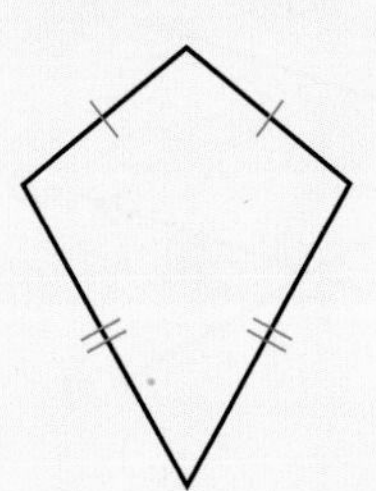

Parallelogram – a quadrilateral with opposite sides equal and parallel

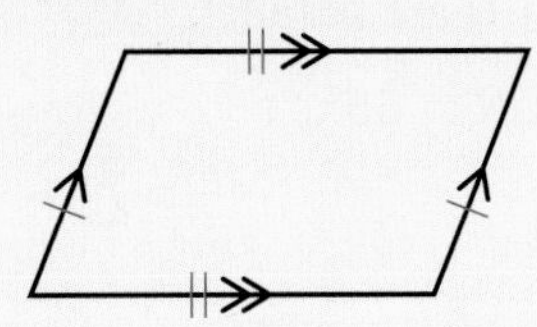

Rhombus – a quadrilateral with four equal sides and opposite sides parallel

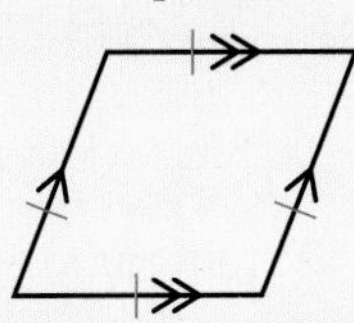

Diagonals of some quadrilaterals have special features. They may be equal, or they may **bisect**, or they may be **perpendicular**.

Square

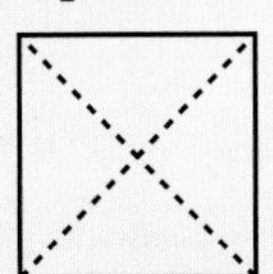

Diagonals equal
Diagonals bisect
Diagonals perpendicular

Rectangle

Diagonals equal
Diagonals bisect

Rhombus

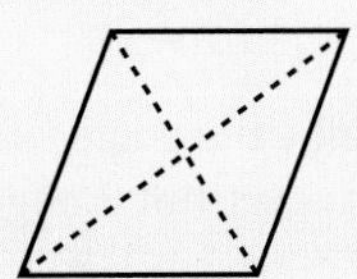

Diagonals bisect
Diagonals perpendicular

Parallelogram

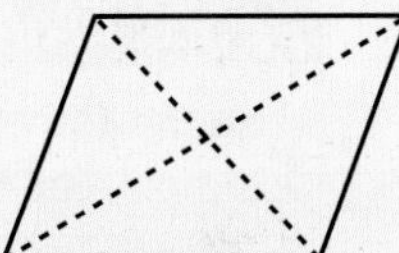

Diagonals bisect

Kite

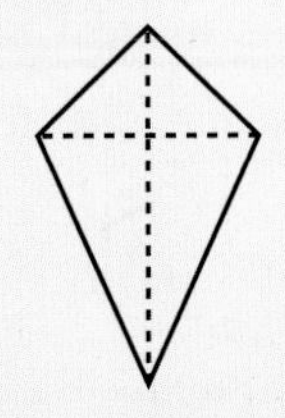

One diagonal bisected
Diagonals perpendicular

Trapezium

Isosceles trapezium

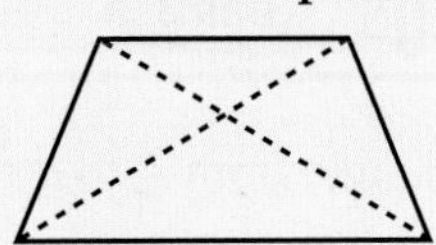

Diagonals equal

Angles in **polygons**:

Polygon	Number of sides	Sum of angles
Triangle	3	180°
Quadrilateral	4	360°
Pentagon	5	540°
Hexagon	6	720°
Heptagon	7	900°
Octagon	8	1080°
Nonagon	9	1260°
Decagon	10	1440°

Bump up your grade

To get a Grade C you need to be able to work out the sum of angles in any polygon. Learn the rule: sum of angles = (number of sides − 2) × 180.

The sum of the angles is always (number of sides − 2) × 180°
In a regular polygon, all angles are equal.
In all polygons, the **exterior angles** (marked opposite) add up to 360°.

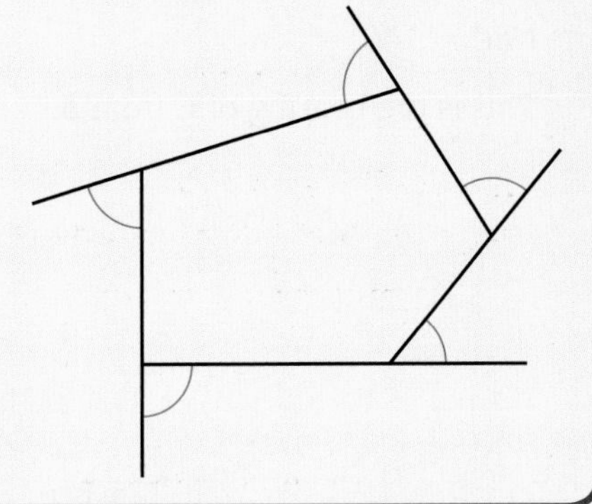

AQA Examiner's tip

Remember that in a regular polygon, all interior angles are equal and all exterior angles are equal.

Example Naming and calculating angles Unit 3

D

In the diagram, AB and CD are parallel.

Angle $AEF = 42°$ and $DGH = 53°$.

a Calculate angles:

i EFG

ii FEG

and give reasons for your answers.

b What type of angle is angle CFE?

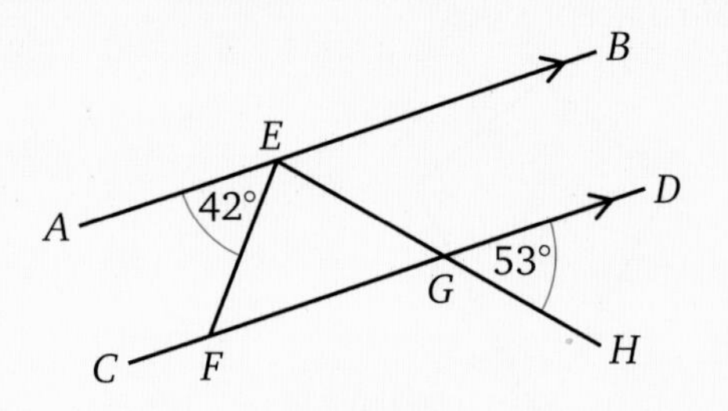

Not drawn accurately

Solution

a **i** $EFG = AEF = 42°$ (alternate angles on parallel lines)

ii $BEG = DGH = 53°$ (corresponding angles on parallel lines)

$FEG = 180 - 42 - 53 = 85°$ (angles on a straight line $= 180°$)

b CFE is obtuse.

AQA Examiner's tip

Use the correct terms to describe the angles. For example, call them alternate angles, not Z angles.

AQA Examiner's tip

When you work out an angle, write it on the diagram as you may need it later.

Example Bearings Unit 3

C

A ship sails in an equilateral triangle from A to B to C.

The bearing of B from A is 020°.

Calculate the bearing of C from B.

Give a reason for each step of your working

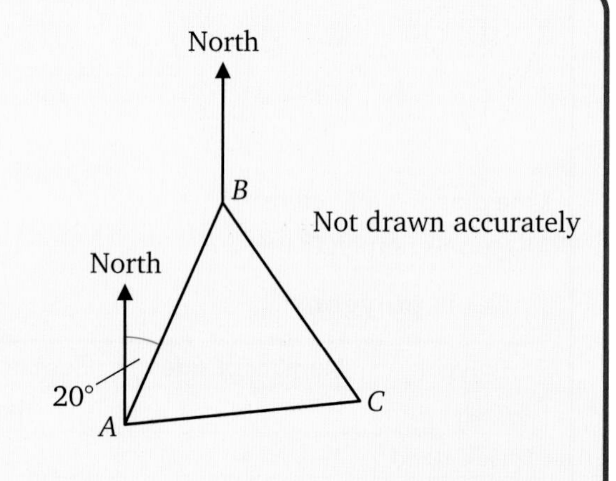

Solution

Angle $ABN = 180 - 20 = 160°$ (interior or allied angles)

Angle $ABC = 60°$ (equilateral triangle)

$NBC = 360 - 60 - 160$ (angles round a point $= 360°$)

$NBC = 140°$

This is the bearing of C from B.

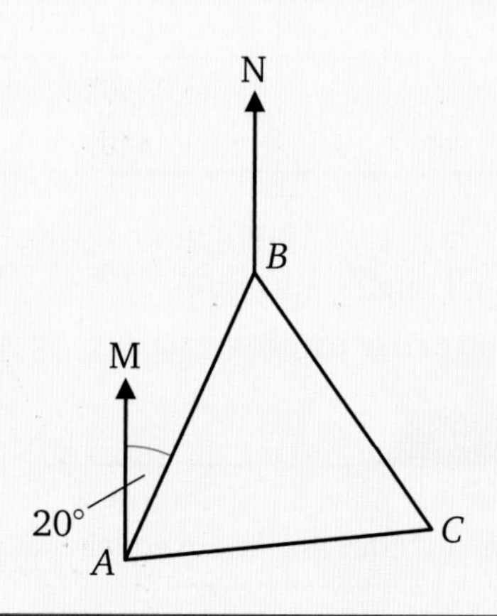

Example Polygons 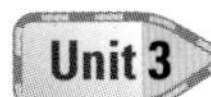

C

a A quadrilateral has two pairs of equal sides. Its diagonals are perpendicular but not equal. What type of quadrilateral is it?

b Calculate the interior angle of a regular decagon.

Solution

a It is a kite.

b **Either** calculate the sum of all 10 angles: $(10 - 2) \times 180 = 1440$
and then divide by 10, to get $1440 \div 10 = 144°$

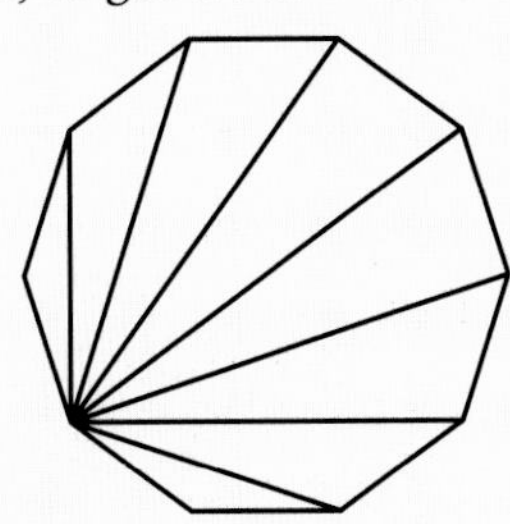

A decagon has 10 sides, and so it is made of 8 triangles

or divide 360 by 10 to get each exterior angle: $360 \div 10 = 36$
and then subtract from 180 to get the interior angle, $180 - 36 = 144°$

Practise... Naming and calculating angles 

G F E D C

G

1 Which angle below is:

a acute

b reflex

c right

d obtuse?

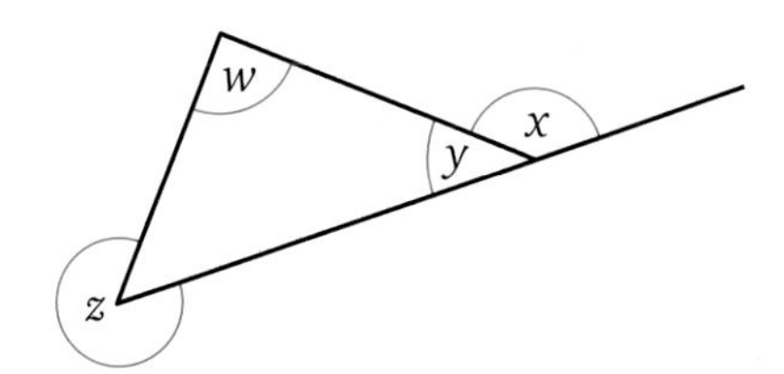

F

2 Calculate the angles a, b and c in the diagram.

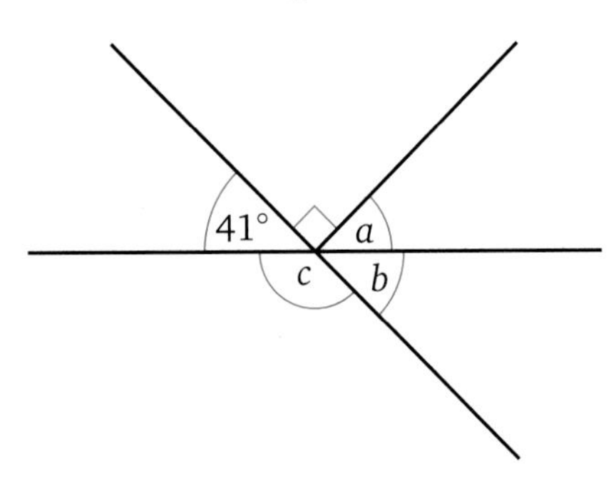

Not drawn accurately

E

3 A quadrilateral has two angles each measuring 100° and one angle of 55°. Calculate the size of the fourth angle.

4 Calculate the size of angle BAC. Give a reason for your answer.

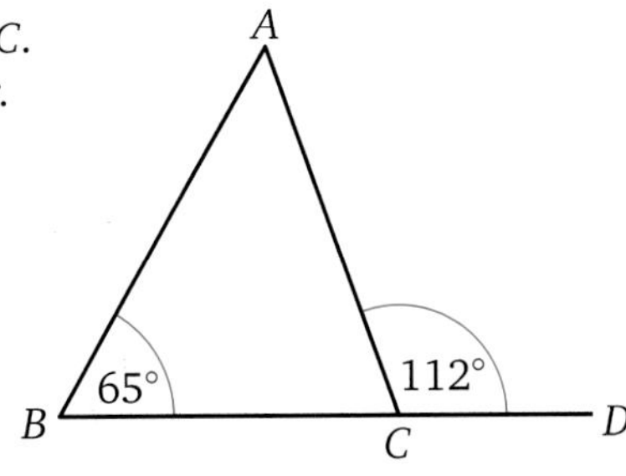

Not drawn accurately

D

5 Measure the bearing of:

a B from A

b C from B

c A from C.

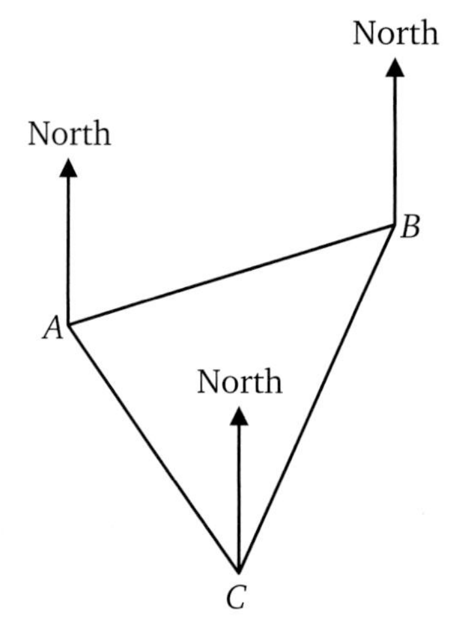

6 AB and CD are parallel.
Find the value of:

a x

b y

c z

Give a reason for each answer.

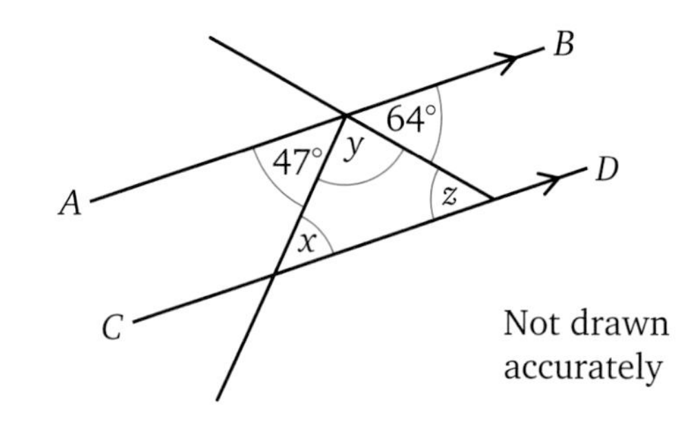

C

7 A regular polygon has 12 sides.
Calculate the size of each exterior angle.

8 A pentagon has three angles of 110°.
The two other angles are equal.
Calculate the size of the two equal angles.

9 The diagram shows an equilateral triangle, a regular hexagon and a regular octagon.
Calculate the size of angle a.

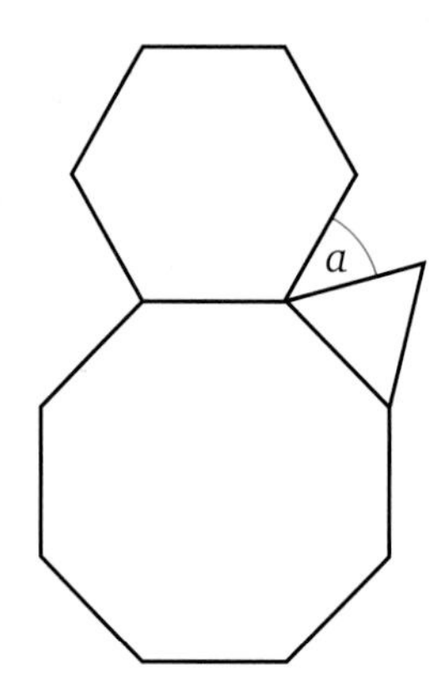

Not drawn accurately

2 Perimeter, area and volume

Unit 3
5 Perimeter and area
11 Area and volume

Key terms

Write down definitions for the following words. Check your answers in the glossary of your Student Book.

area
base
circle
circumference
cross-section
cube
cuboid
cylinder
diameter
net
perimeter
perpendicular height
prism
radius
surface area
triangular prism
volume

Revise... Key points

Perimeter and area Unit 3

The **perimeter** of a shape is the distance around it.

The **area** is the amount of space inside the shape.

You can find the perimeter of this shape by counting the 1 cm lines around a shape (see the green numbers outside the rectangle).

The perimeter is 10 cm

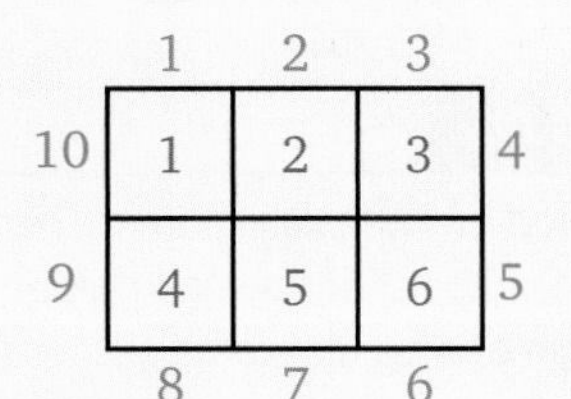

The area is 6 cm^2

You can find the area by counting the squares inside a shape (see the blue numbers inside the rectangle).

You can estimate the area of irregular shapes by counting the squares inside. Count any square that is more than half inside the shape, and ignore those that are less than half inside the shape.

Area of quadrilaterals and triangles

The area of a rectangle, parallelogram and trapezium are all calculated by multiplying two perpendicular measurements.

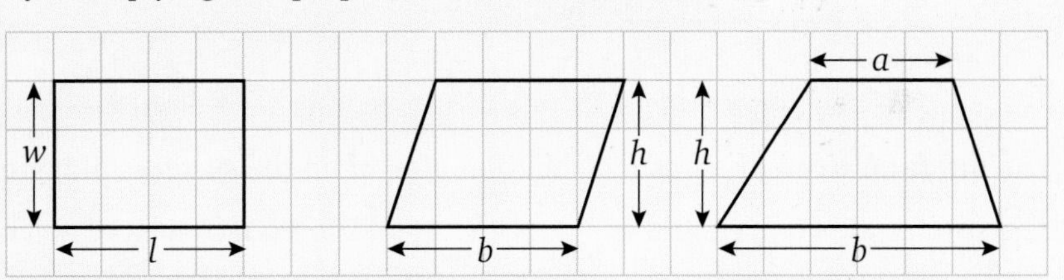

Rectangle:
Area = length × width
$A = lw$

Parallelogram:
Area = **base** × height
$A = bh$

Trapezium:
Area = average width × height
$A = \frac{1}{2}(a + b)h$

AQA Examiner's tip

The formula for the area of a trapezium is given on the exam paper, so you do not need to learn it. But you must learn the other formulae.

A triangle is half a parallelogram.

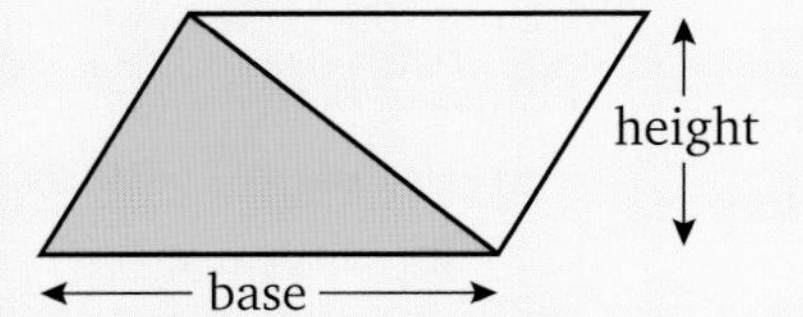

AQA Examiner's tip

Remember: in all of these formulae, the height and base are perpendicular.

So the area of a triangle is $\frac{1}{2}$ × base × height or $A = \frac{1}{2}bh$

Circles Unit 3

Circumference = $\pi \times$ **diameter**, or $C = \pi d$

Area = $\pi \times$ (**radius**)^2, or $A = \pi r^2$

π is approximately 3.14. Many calculators have a π key that you should use, as it gives greater accuracy.

Diameter = $2 \times$ radius, or $d = 2r$

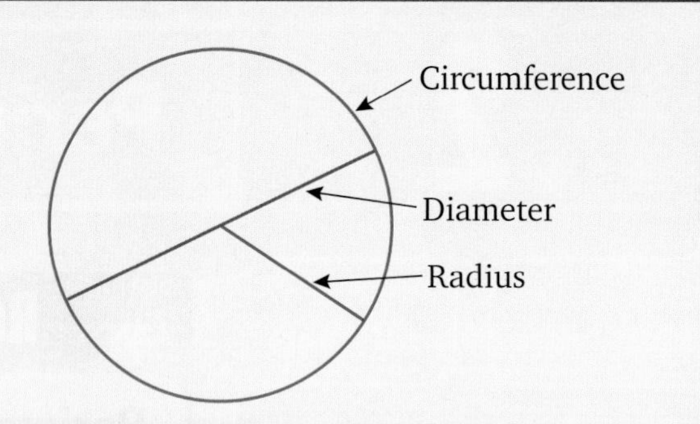

Prisms Unit 3

A **prism** is a shape with the same **cross-section** all the way through.

The formula for calculating the **volume** of a prism is:

Volume = Area of cross-section $\times$ length

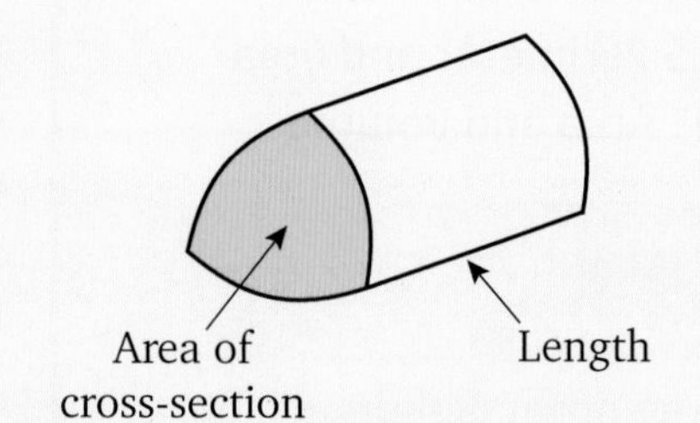

This formula is included on the exam paper. The area of cross-section is the shape that goes all through the solid, marked in orange on this diagram.

If the prism is stood on its end, then the formula for volume becomes:

Volume = area of cross-section $\times$ height

The volume of a **cuboid** = Area of top $\times$ height

$V = l \times w \times h$

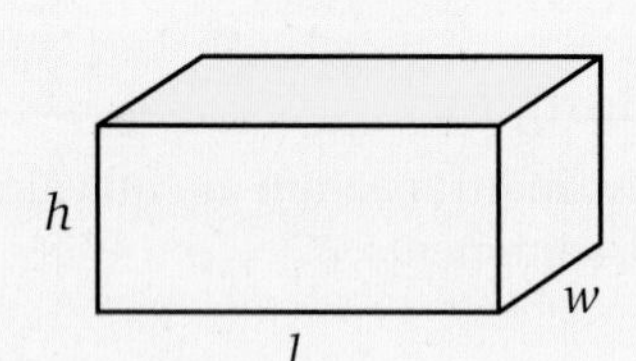

The volume of a **cylinder** = Area of top $\times$ height

$V = \pi r^2 \times h$

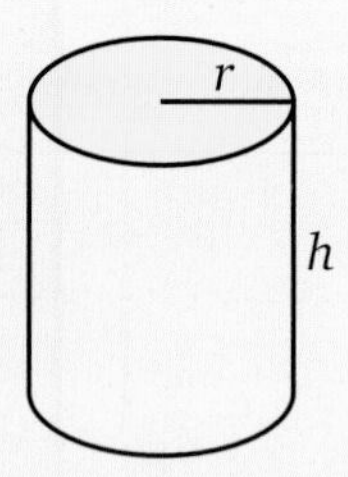

The **surface area** of a prism is the total area of all the surfaces. Imagine the **net** so you include all surfaces.

The surface area of a cuboid is:

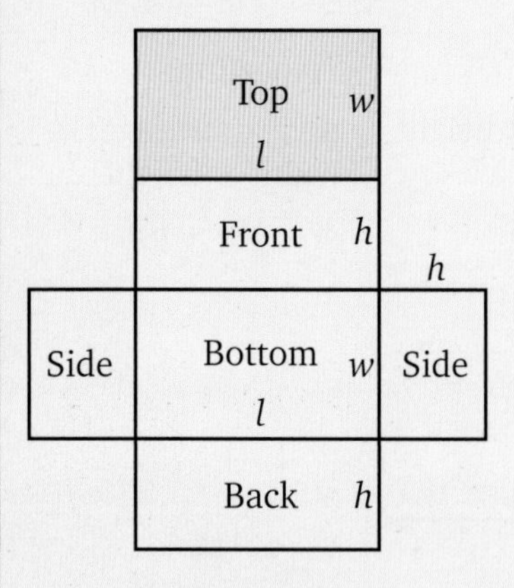

Area of top (lw)
+ Area of bottom (lw)
+ Area of front (lh)
+ Area of back (lh)
+ Area of right side (wh)
+ Area of left side (wh)
Surface area = $2lw + 2lh + 2wh$

The surface area of a cylinder is:

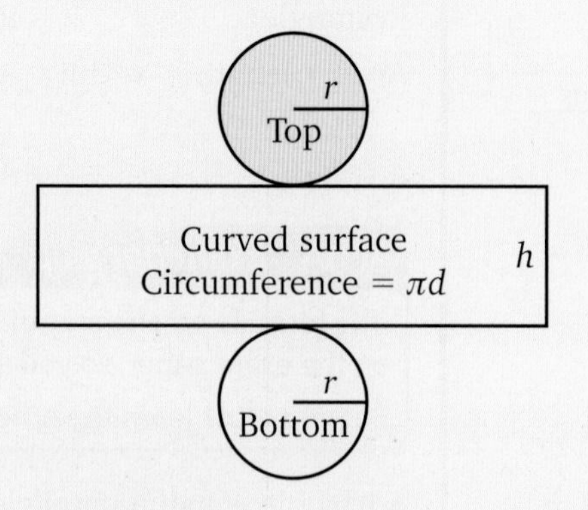

Area of top (πr^2)
+ Area of bottom (πr^2)
+ Area of curved surface

The curved surface can be unrolled to make a rectangle with a length equal to the circumference and a height of h.

So the area of curved surface = Circumference $\times h = \pi dh$

Surface area = $2\pi r^2 + \pi dh$

Example Perimeter and area 

a Find the perimeter and area of the shape below.

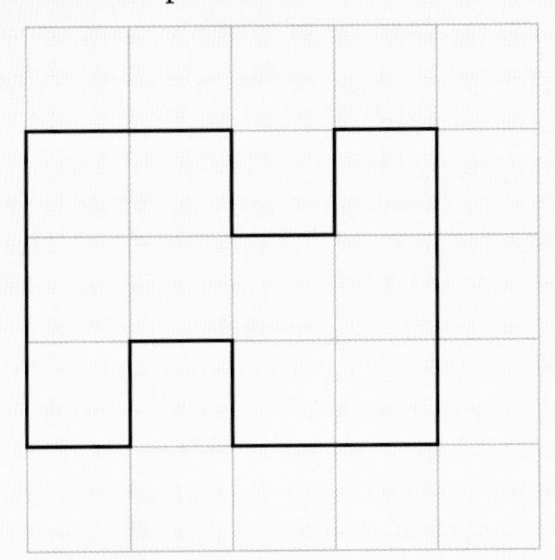

b Estimate the area of the shape below.

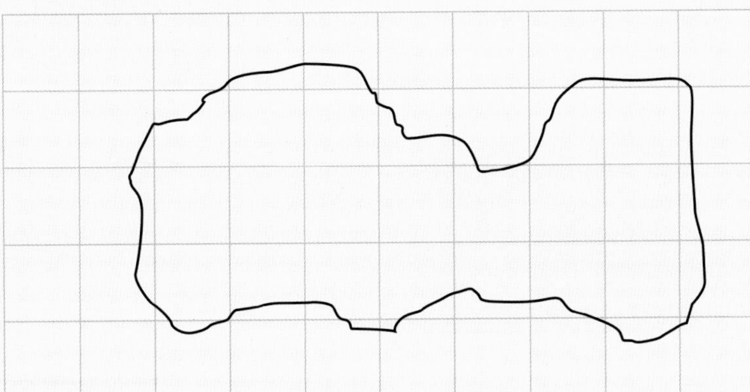

Solution

a The perimeter is the number of centimetres around the outside, which is 18 cm.

The area is the number of squares inside, which is 10 cm^2.

b By counting the squares and those more than half inside the shape, the area is approximately 19 cm^2.

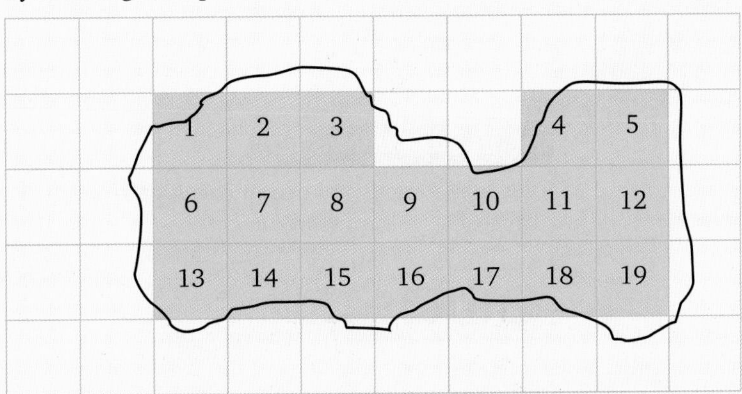

Example Area of shape made from quadrilaterals and triangles Unit 3

D

Calculate the area of this shape.

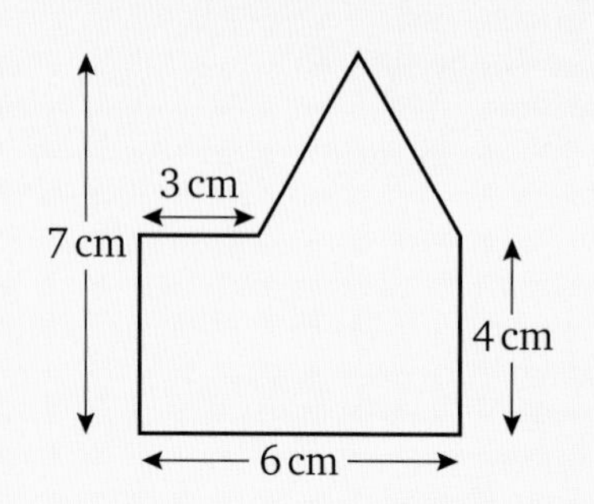

Not drawn accurately

Solution

The shape can be split into a rectangle and a triangle.

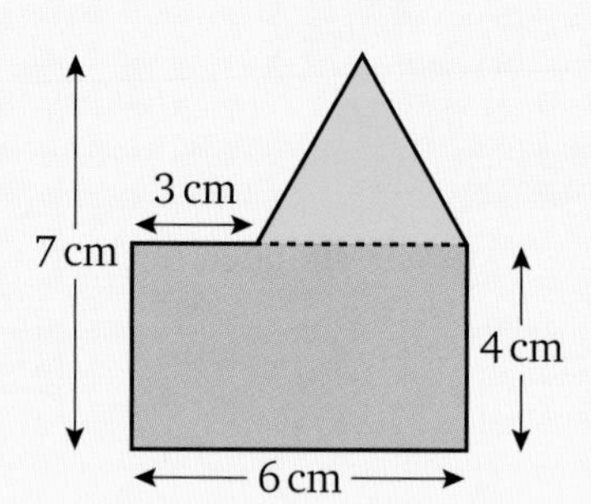

Area of rectangle $= lw = 6\text{ cm} \times 4\text{ cm} = 24\text{ cm}^2$

Base of triangle $= 6\text{ cm} - 3\text{ cm} = 3\text{ cm}$

Height of triangle $= 7\text{ cm} - 4\text{ cm} = 3\text{ cm}$

Area of triangle $= \frac{1}{2}bh = \frac{1}{2} \times 3 \times 3 = 4.5\text{ cm}^2$

Total area $= 24\text{ cm}^2 + 4.5\text{ cm}^2 = 28.5\text{ cm}^2$

AQA Examiner's tip

Do not be afraid to draw on the diagrams on the exam paper. Annotate them with any measurements that you work out.

Example Area and circumference of circles Unit 3

D

A **circle** has a radius of 8 cm.

Calculate its circumference and area.

Solution

Circumference $= \pi \times$ diameter

Circumference $= \pi \times 16$ because diameter $= 2 \times$ radius

Circumference $= 50.2654825\ldots$

Circumference $= 50.3$ cm (to 1 decimal place)

AQA Examiner's tip

Make sure you include units in your answer. Circumference is a length, and so it is measured in cm, mm or m. Area is measured in cm^2, mm^2 or m^2.

Area $= \pi r^2 = \pi \times 8^2 = 201.0619298\ldots$

Area $= 201.1\text{ cm}^2$ (to 1 decimal place)

AQA Examiner's tip

Do not mix up the two formulae. Area $= \pi r^2$, because area is measured in cm^2.

Key it straight into your calculator:

Example Surface area and volume Unit 3

D

This prism is made from **cubes** with sides of 2 cm.

Find the volume and surface area of the prism.

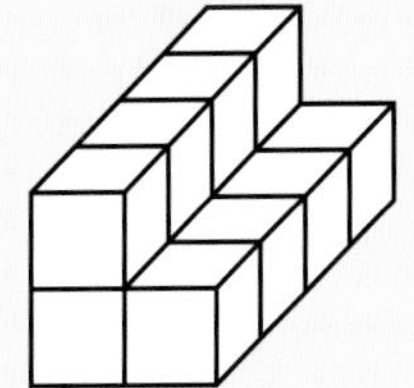

Solution

Volume:

Area of front $= 3 \times 2^2 = 12\,\text{cm}^2$

Volume $=$ area $\times$ length $= 12 \times 8 = 96\,\text{cm}^3$

Surface area:

Area of front $= 12\,\text{cm}^2$

Area of back $= 12\,\text{cm}^2$

Area of bottom $= 4 \times 8 = 32\,\text{cm}^2$

Area of left side $= 32\,\text{cm}^2$

4 other faces each have an area of $2 \times 8 = 16\,\text{cm}^2$

Total surface area $= 2 \times 12 + 2 \times 32 + 4 \times 16 = 152\,\text{cm}^2$

AQA Examiner's tip

Do not forget to include the correct units: cm^2 for area, cm^3 for volume.

Practise... Area and volume

1 Find the area and perimeter of the shape below.

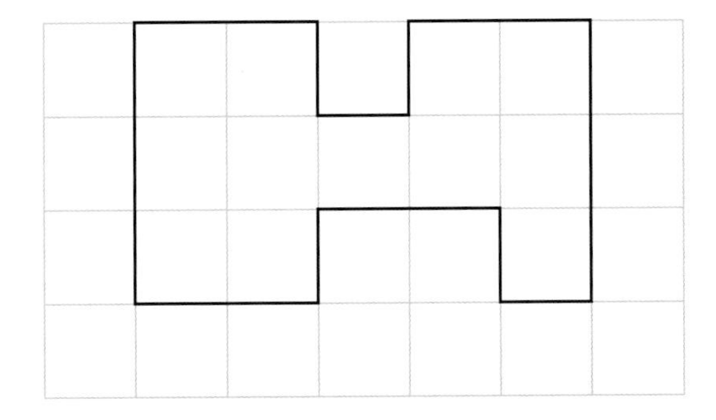

2 Estimate the area of the shape below.

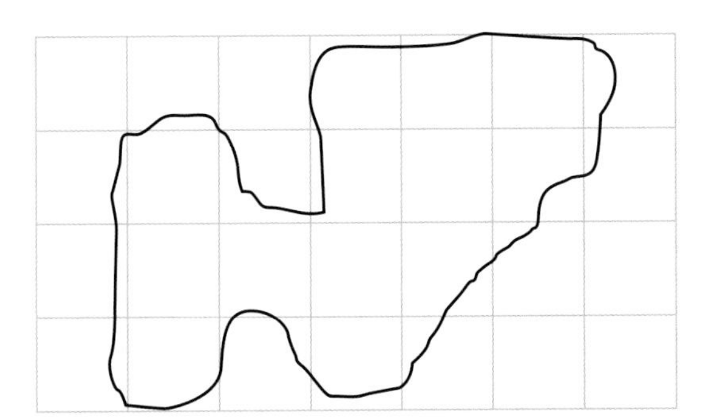

F

3 A rectangle is 6 cm long and 4 cm wide.
Calculate the area and perimeter.

E

4 A rectangle has a length of 7.2 cm and a width of 4.3 cm.
Calculate the area and perimeter.

D

5 Calculate the area of these shapes:

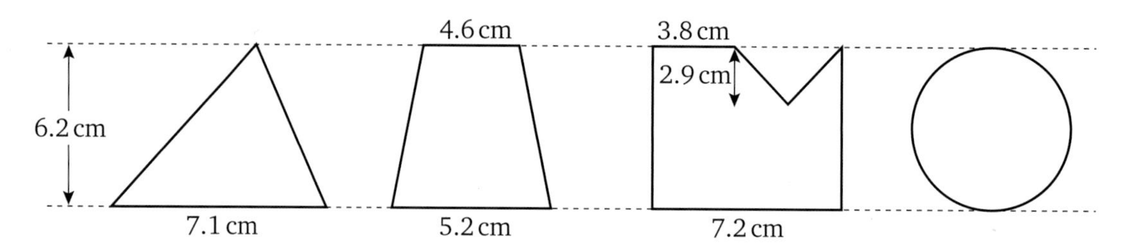

6 Find:

a the volume

b the surface area

of this cuboid.

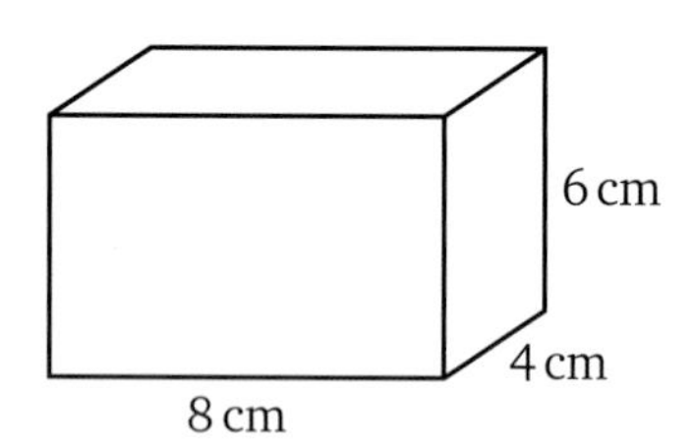

C

7 A cylinder has a radius of 4 cm and a height of 6 cm.
Calculate the volume of the cylinder.

8 This cuboid is 12 cm long and 2 cm high.
Its volume is 96 cm^3.

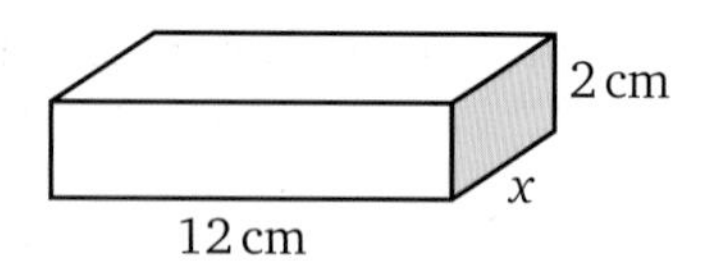

Calculate its width, x.

Geometry and measure

3 Transformations

Unit 3

9 Reflections, rotations and translations
14 Enlargements

Key terms

Write down definitions for the following words. Check your answers in the glossary of your Student Book.

angle of rotation
centre of enlargement
centre of rotation
coordinates
enlargement
line of symmetry
order of rotation symmetry
reflection
rotation
scale factor
similar
transformation
translation
vector
vertex (pl. vertices)

Revise... Key points

Reflection Unit 3

Reflection symmetry is where half a shape is a mirror image of the other half. The **mirror line** is called the **line of symmetry**.

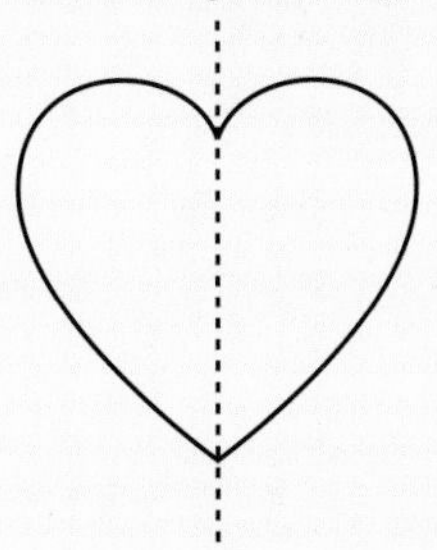

A reflection is a **transformation** where a shape gets reflected in a mirror line.

Rotation Unit 3

The **order of rotation symmetry** is the number of different positions a shape can turn round into so that it occupies the same space.

This shape has **rotation** symmetry of order 4.

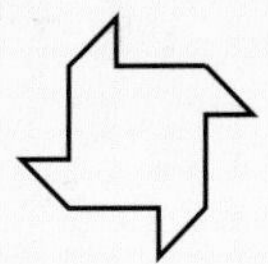

A rotation is a transformation where a shape gets turned or rotated about a fixed point.

To rotate a shape, you need to know the **centre of rotation**, the angle to turn and the direction.

Translation Unit 3

A **translation** is a transformation where every point moves the same distance in the same direction.

It is described by a **vector**.

The vector $\begin{pmatrix} 3 \\ -2 \end{pmatrix}$ means move 3 to the right (positive = right, negative = left)
2 down (positive = up, negative = down)

Enlargements Unit 3

An **enlargement** changes the size of an object, but it keeps the same shape.

The lengths of all the sides of the original object are multiplied by the same amount. This amount is called the **scale factor**.

The distance of the object from the **centre of enlargement** is also multiplied by the scale factor.

Example Reflection Unit 3

D

Reflect triangle A in the line $x = 2$

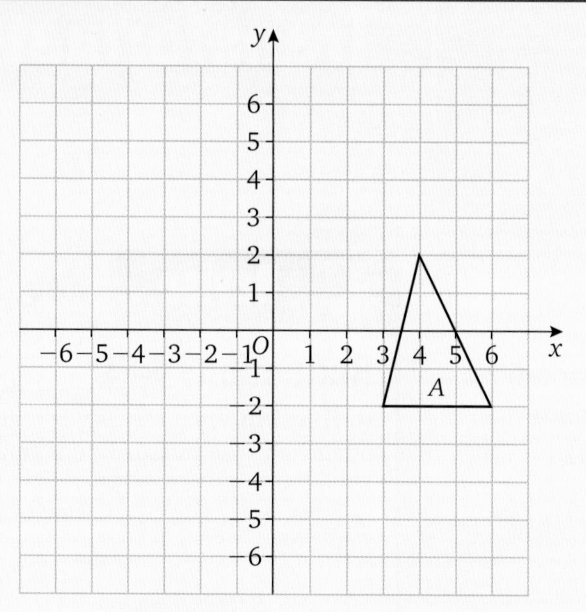

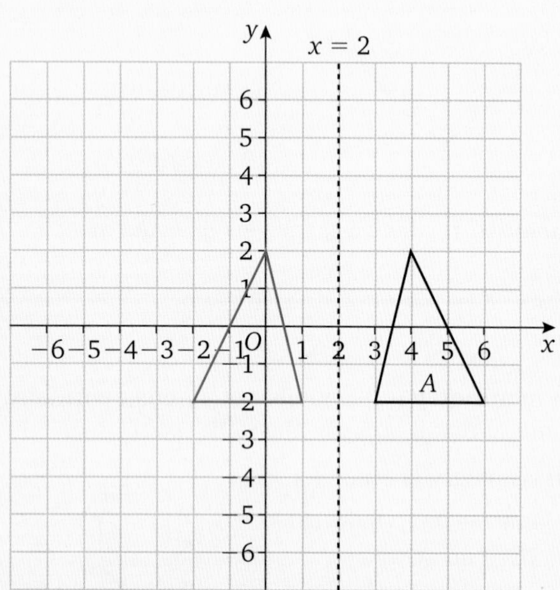

Solution

Remember: the distances from the mirror line stay the same.

For example, the top **vertex** of triangle A is 2 squares to the right of the mirror line, so the reflection is 2 squares to the left of the mirror line.

AQA Examiner's tip

Make sure you know that mirror lines like $x = 3$ are vertical, passing through 3 on the x-axis, and lines like $y = -2$ are horizontal, passing through -2 on the y-axis.

Bump up your grade

For a Grade C, you need to recognise the lines $y = x$ and $y = -x$, and reflect shapes in them.

Example Rotation Unit 3

D

Rotate shape A 90° anticlockwise about the origin.

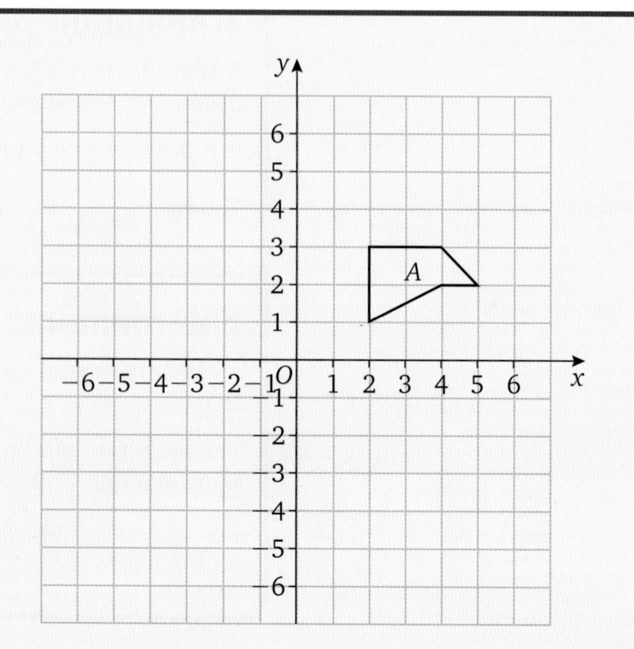

Solution

Use tracing paper to trace the shape. Hold your pencil still at (0, 0), which is the origin.

Rotate the tracing paper 90° anticlockwise (the opposite direction to the hands on a clock).

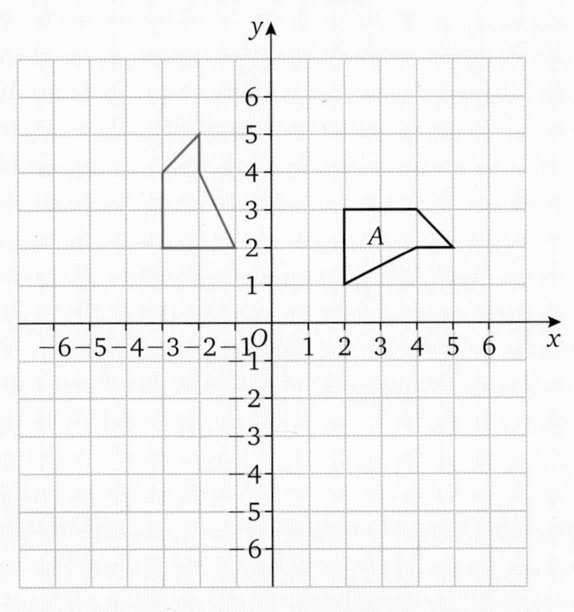

Trace the shape back on to the question paper.

Example Translation Unit 3

Describe the translation that maps:

a A to B

b B to A

c A to C

d C to A.

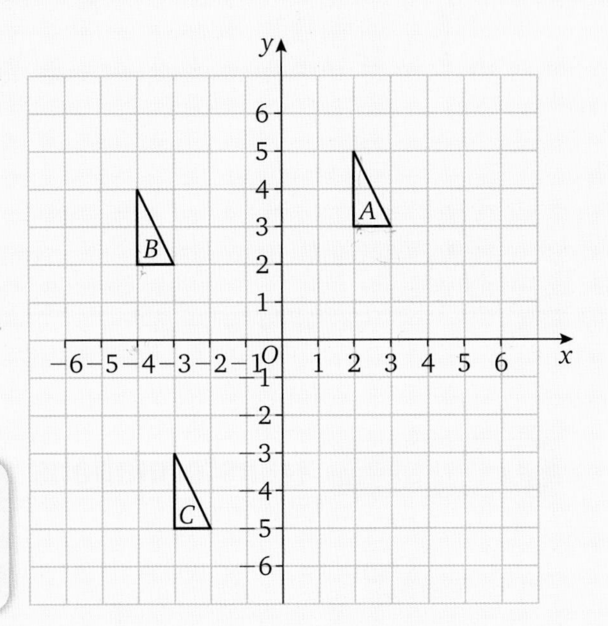

Solution

a $\begin{pmatrix}-6\\-1\end{pmatrix}$

b $\begin{pmatrix}6\\1\end{pmatrix}$

c $\begin{pmatrix}-5\\-8\end{pmatrix}$

d $\begin{pmatrix}5\\8\end{pmatrix}$

Hint

The translation from B to A is the opposite of A to B, and C to A is the opposite of A to C, so the signs all change to the opposite direction.

Bump up your grade

You could describe **a** as '6 to the left and 1 down', but for a Grade C you should use column vectors.

C

Example Enlargements Unit 3

D

Enlarge the trapezium with a scale factor of 3, centre $(-4, 2)$.

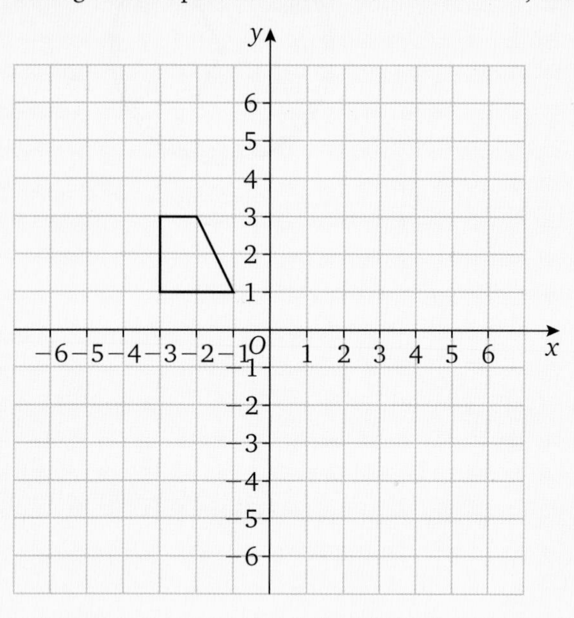

Solution

Mark the centre of enlargement.

Draw lines from the centre through each vertex and extend them to three times the length.

Join the ends of these lines to produce the enlarged shape.

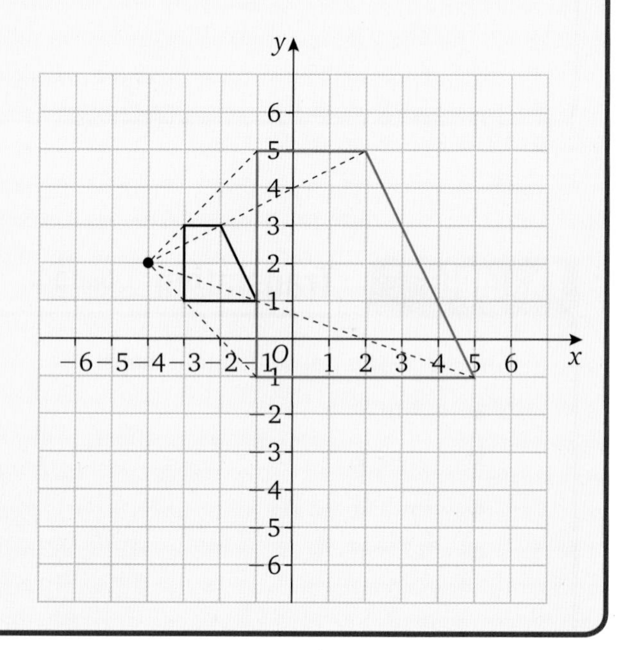

Practise... Transformations Unit 3

F E D C

F

1 Copy these shapes and draw any lines of reflection symmetry.
Write down the order of rotation symmetry of each shape.

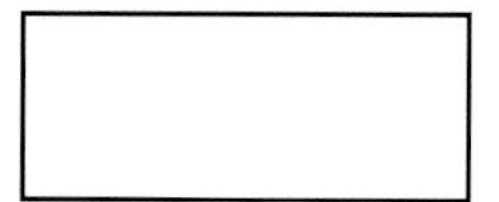

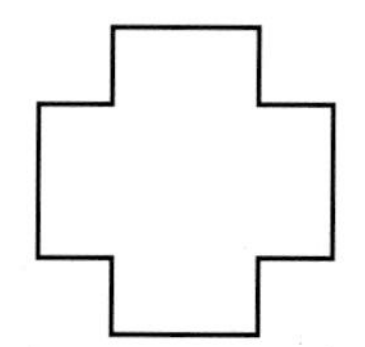

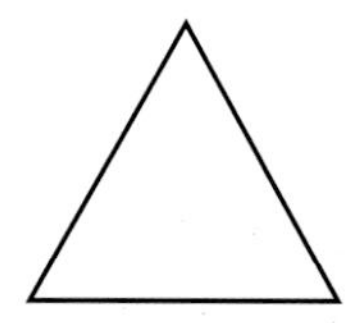

Questions 2–4 are about the diagram on the right.

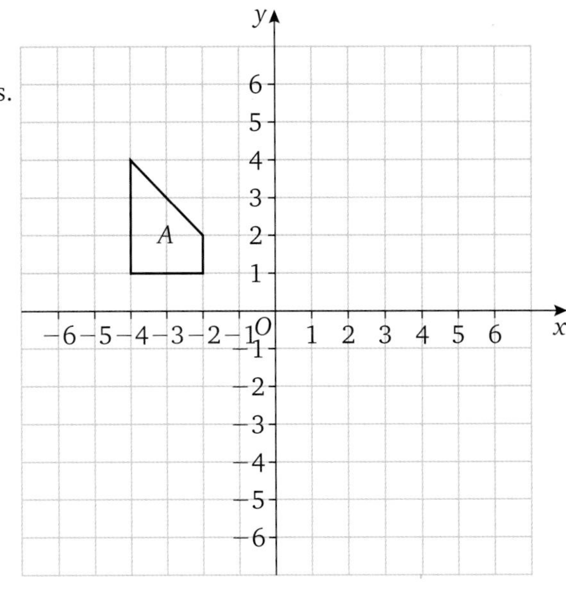

2 Copy the diagram and reflect shape A in the x axis.

E

3 Rotate shape A 90° clockwise about the origin.

D

4 Reflect shape A in the line $y = -x$.

C

5 Copy the diagram and shape A.

a Draw an enlargement of A, using a scale factor of 2 and a centre of enlargement (0, 0).

b Translate A through the vector $\begin{pmatrix} 4 \\ -4 \end{pmatrix}$.

c Rotate A 180° anticlockwise about the point (1, 0).

D

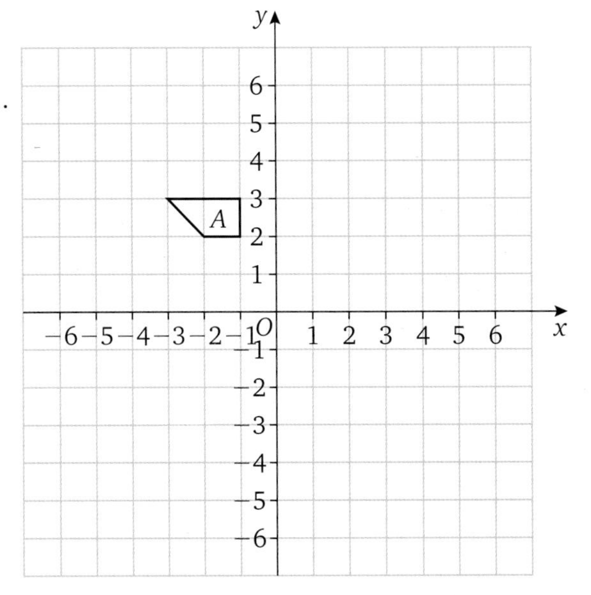

6 Describe fully the transformation that maps shape:

a A onto B

b C onto E

c A onto C

d C onto D

e C onto B.

C

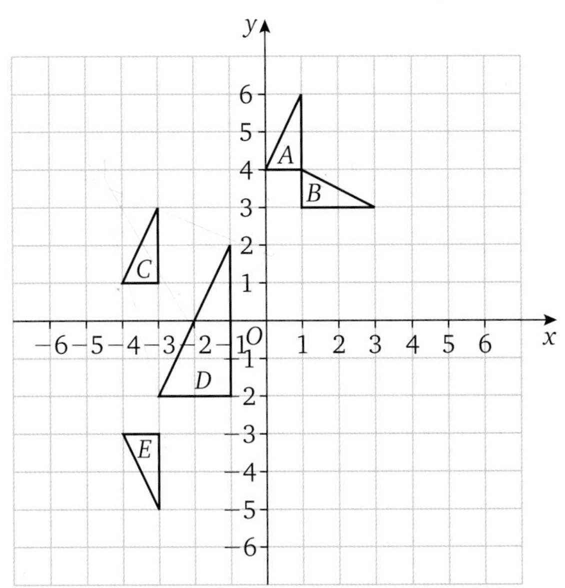

7 **a** Draw a rectangle with a length of 4 cm and a width of 2 cm.

b Draw an enlargement of your rectangle, so that the length is 10 cm.

Geometry and measure

4 Measures, loci and construction

Unit 3

Key terms

Write down definitions for the following words. Check your answers in the glossary of your Student Book.

arc
bisect
bisector
capacity
compound measure
congruent
construction (construct)
conversion factor
edge
elevation
equidistant
equilateral triangle
face
locus (pl. loci)
mass
net
perpendicular
plan
scale
similar
unit
vertex (pl. vertices)

Look back at Chapter 15 (Construction), to remind yourself about similarity and congruence.

Revise... Key points

Measures Unit 3

The common metric **units** are:

length	**mass**	**capacity**	
10 mm = 1 cm	1000 mg = 1 g	10 ml = 1 cl	1 ml = 1 cm^3
100 cm = 1 m	1000 g = 1 kg	100 cl = 1 l	
So 1000 mm = 1 m	1000 kg = 1 tonne	1000 ml = 1 l	1 l = 1000 cm^3
1000 m = 1 km			

You need to know approximate equivalents for some imperial units:

5 miles ≈ 8 kilometres
2.2 pounds ≈ 1 kilogram
4.5 litres ≈ 1 gallon
1 inch ≈ 2.5 centimetres

AQA Examiner's tip

Remember that time is not metric. For example, 90 seconds is 1 minute 30 seconds. This is not 1.30 minutes. It is 1.5 minutes.

Compound measures Unit 3

Compound measures combine two units.

Use the units to tell you what calculation to do.

To calculate a speed in km/h, remember km/h means $\frac{\text{km}}{\text{h}}$ or km ÷ hours.

Fuel consumption in km/l means $\frac{\text{km}}{l}$ or km ÷ litres used.

Constructing triangles Unit 3

Given a side and 2 angles

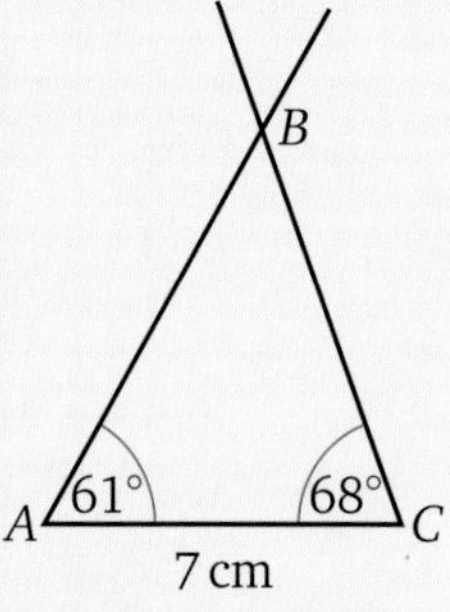

Draw AC 7 cm long
Measure angle BAC = 61°
Draw a long line
Measure angle BCA = 68°
Draw a long line to cross at B

Given 2 sides and an angle

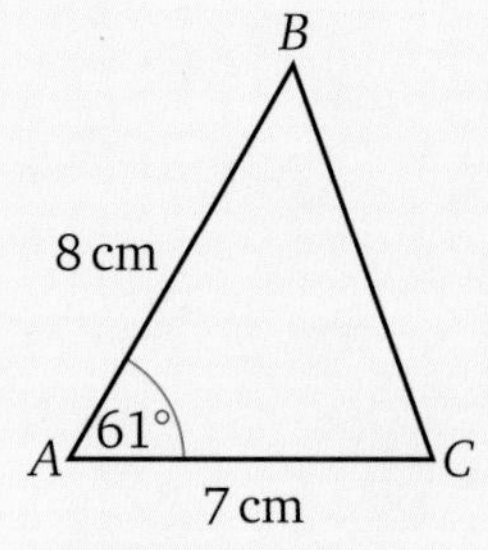

Draw AC 7 cm long
Measure angle BAC = 61°
Measure AB = 8 cm
Join B to C

Given 3 sides.

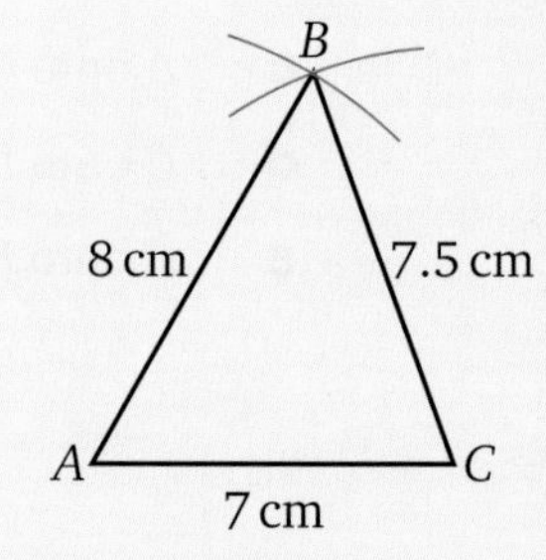

Draw AC 7 cm long
Open compasses to 8 cm
With compass point on A, draw an **arc**
Open compasses to 7.5 cm
The arcs intersect at B
With compass point on C, draw an arc
The arcs intersect at B

AQA Examiner's tip

Remember to leave your arcs as they are your working out, and they show your method.

Constructions and loci Unit 3

A **locus** is a collection of points that meets a condition.

A **perpendicular bisector** bisects a line (cuts it exactly in half) at right angles.

To **bisect** *AB*:

Open your compasses more than half the length of *AB*.

With the compass point on *A*, draw arcs on either side of the line *AB*.

With the compass point on B, draw arcs on either side of the line *AB*.

The arcs meet at *C* and *D*.

CD is the perpendicular bisector of *AB*.

CD is the locus of points **equidistant** (the same distance) from *A* and *B*.

C
A B
D

An angle bisector bisects an angle.

To bisect angle *BAC*:

Put the compass point on *A* and make equal arcs to cross *AB* and *AC* at *X* and *Y*.

Make two more arcs, with *X* and *Y* as centres, to meet at *Z*.

AZ bisects angle *BAC*.

AZ is the locus of points equidistant from *AB* and *AC*.

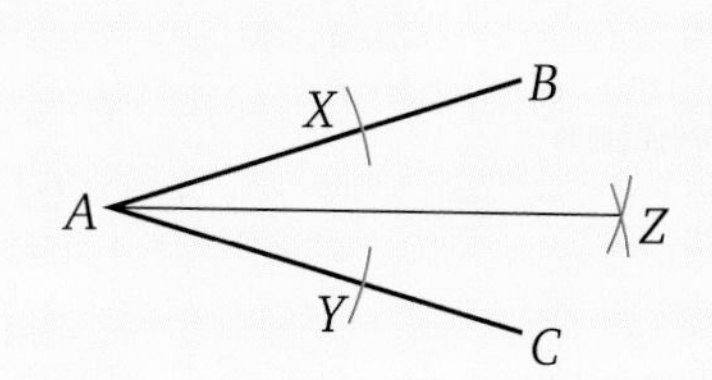

Bump up your grade

At Grade C you need to know that the perpendicular bisector of *AB* joins all of the points equidistant from *A* and *B*, and that the bisector of angle *ABC* joins all points equidistant from *BA* and *BC*.

You might also need to use this to solve problems.

Three-dimensional shapes Unit 3

Here is a cuboid. It is 4 cm long, 1 cm wide and 3 cm high.

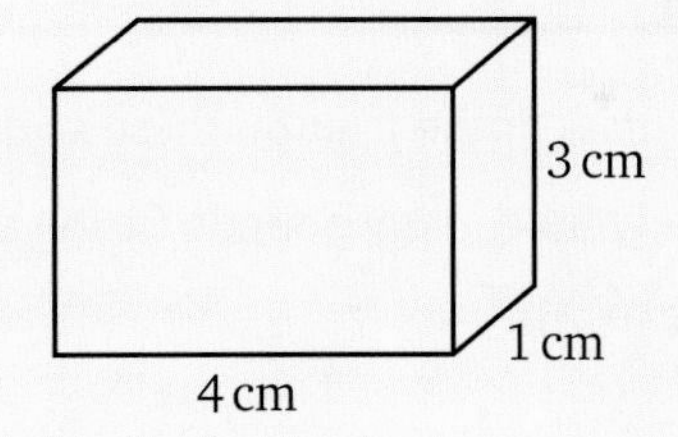

Its **net** might look like this:

4 cm
3 cm
1 cm
3 cm
1 cm

An isometric drawing of the cuboid would look like this:

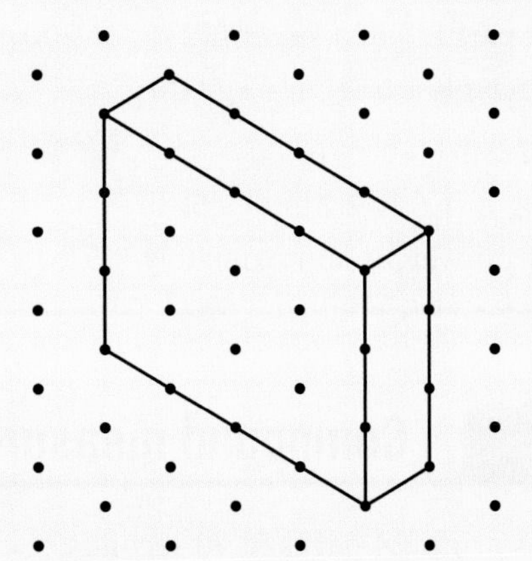

Notice that isometric drawings rarely have horizontal lines.

It usually helps to draw the front face first.

Hint

Opposite **faces** are the same size.

Edges that will meet must be the same length.

AQA Examiner's tip

Isometric (dotty) paper must be the right way round.

The vertical dots are closer together than the horizontal ones.

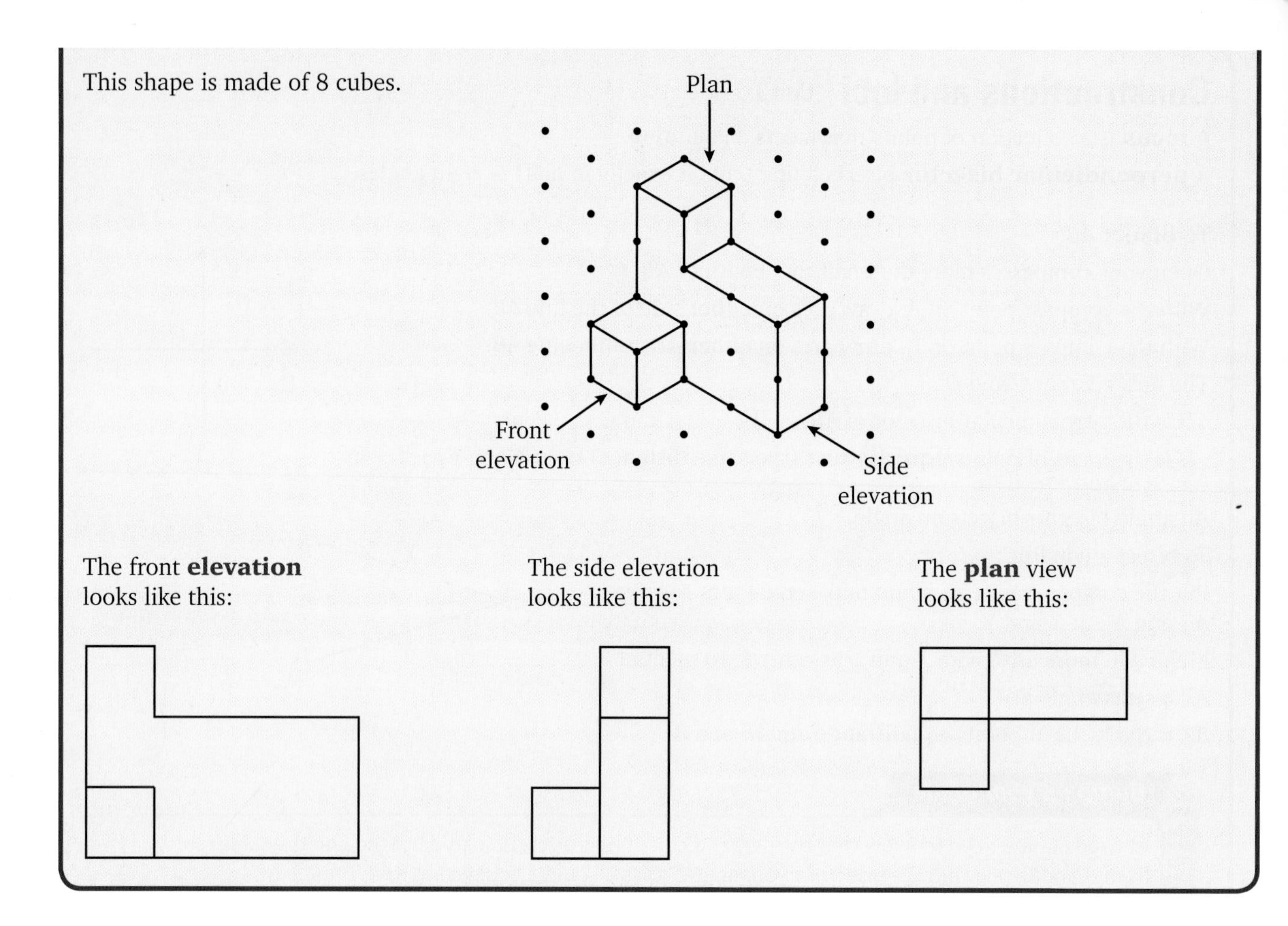

Example Measures 

a What is the mass of the fruit on these scales?

b Approximately how many pounds (lb) is this?

Solution

a Each kilogram is split into 5 parts.
So each part is $\frac{1}{5}$ of a kg, or 0.2 kg, or 200 g.
So the mass is 1.6 kg, or 1 kg 600 g.

b $1\text{ kg} \cong 2.2\text{ lb}$
So $1.6\text{ kg} \cong 1.6 \times 2.2\text{ lb} = 3.52\text{ lb}$
So the fruit weighs approximately 3.5 lb.

Example Compound measures 

Terry's printer prints 24 pages in 90 seconds.
What is the print speed in pages/minute?

Solution

Pages/minute means $\frac{\text{pages}}{\text{minutes}}$ or pages ÷ minutes.
90 seconds = 1.5 minutes.

So the print speed = 24 ÷ 1.5 = 16 pages/minute

AQA Examiner's tip
Make sure you use the correct units. In this example you need to change seconds into minutes.

Example Constructions and loci Unit 3

C

A triangular field *ABC* has sides of 120 m, 90 m and 100 m as shown.

A telegraph pole is to be placed closer to *AC* than *AB*, and more than 50 m from *A*.

Use a scale of 1 cm to 10 m to make a scale drawing and **construct** the possible position of the pole.

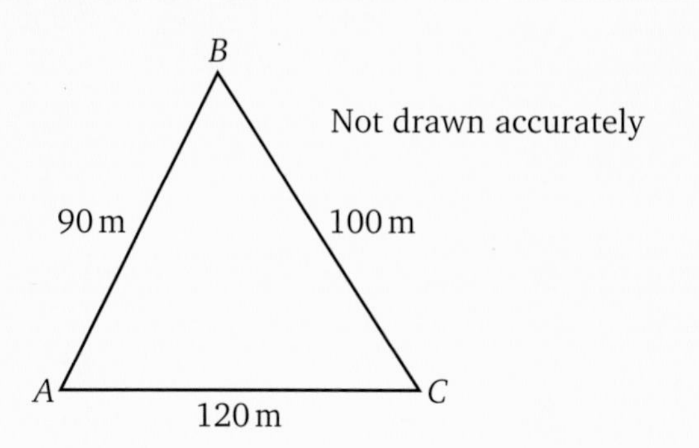

Solution

Use compasses to construct a triangle of sides 12 cm, 9 cm and 10 cm.

Construct the perpendicular bisector of angle *BAC*.

Construct the locus of points 5 cm from *A*.

Shade the region where the pole can go.

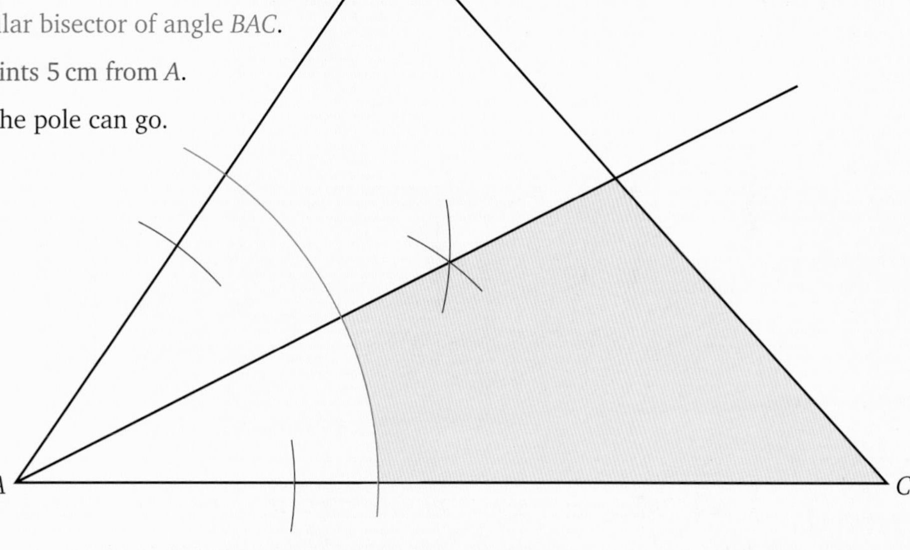

Practise... Measures, loci and constructions Unit 3

G F E D C

G

1 Here is a net.

a What is the name of the three-dimensional shape it will make?

b Which **vertices** will meet with *A* when it is made?

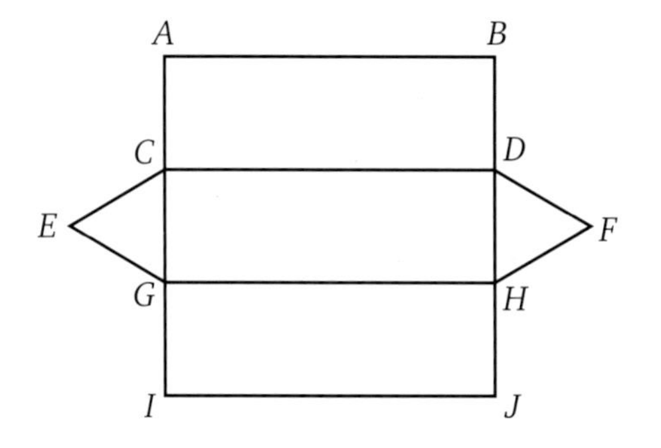

F

2 **a** A door is 190 cm high. Write 190 cm in metres.

b Roughly how many feet is this?

E

3 Draw the net of a cuboid that is 4 cm long, 2 cm wide and 1 cm high.

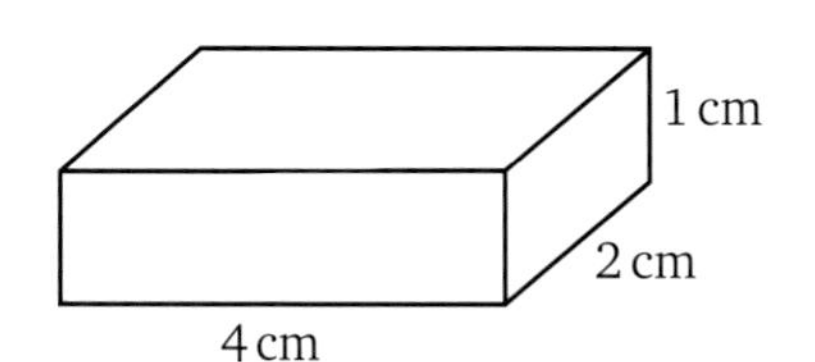

D

4 This prism has a front elevation in the shape of an **equilateral triangle** with sides of 6 cm.
Use ruler and compasses to make an accurate construction of the front elevation.

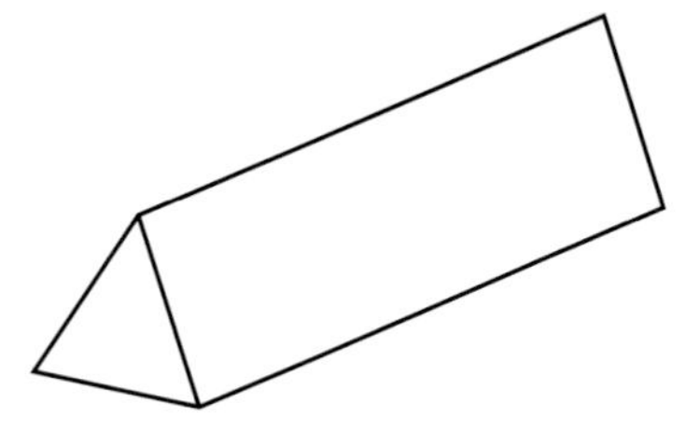

5 Jamie takes 12.6 seconds to run 100 m.
Calculate his speed in m/s.

6 Look at this plan view of a solid.

Which one of the three shapes below could it NOT be?

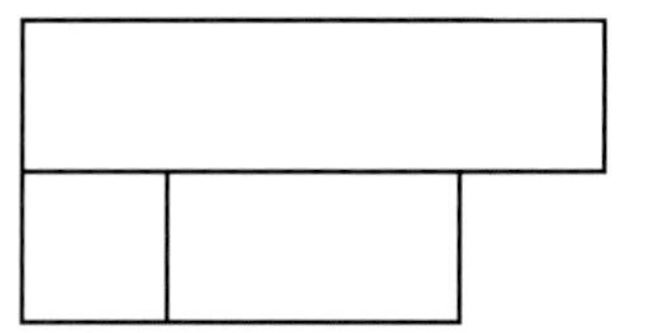

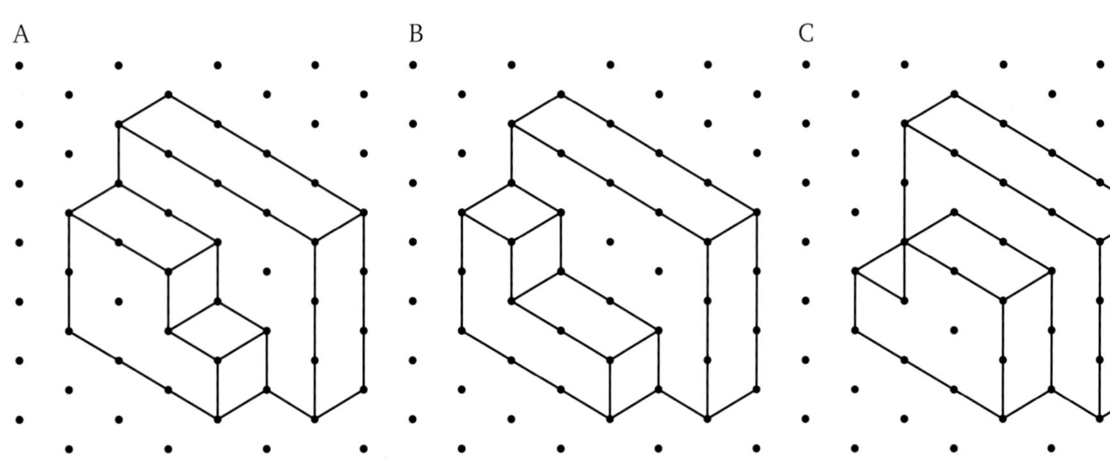

C

7 **a** Construct triangle ABC with $AB = 8$ cm, $AC = 9$ cm and angle $BAC = 72°$.

b Construct the locus of points equidistant from A and C.

8 **a** A rectangular field $ABCD$ is 120 m long and 80 m wide.
Construct an accurate drawing of the field, using a scale of 1 cm to 10 m.

b Two scarecrows are placed in the field, so that one is 60 m from corners A and D, and the other is 60 m from B and C.
Construct the positions of the scarecrows. Label them S and T.

c By taking measurements from your diagram, calculate the distance between the two scarecrows.

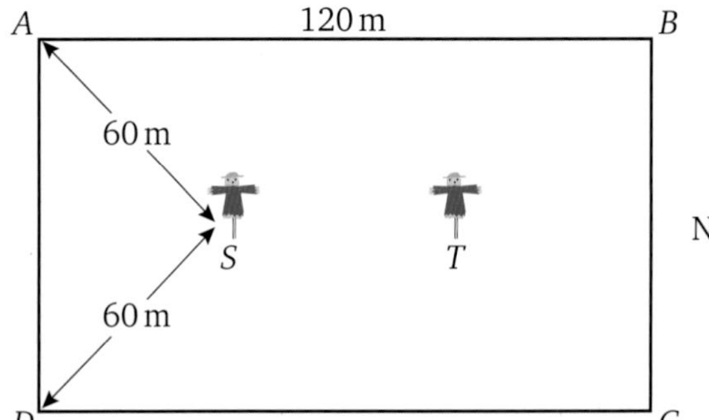

Not drawn accurately

9 Hugo cycles to his friend's house, 9 km away, in 12 km/h.

a Calculate how long it takes him.

b His journey back home takes him 54 minutes. Calculate his speed home.

Geometry and measure

5 Pythagoras' theorem

Unit 3

18 Pythagoras' theorem

Key terms

Write down definitions for the following words. Check your answers in the glossary of your Student Book.

hypotenuse

Pythagoras' theorem

Revise... Key points

Pythagoras' theorem Unit 3

In any right-angled triangle the longest side is always opposite the right angle.

This side is called the **hypotenuse**.

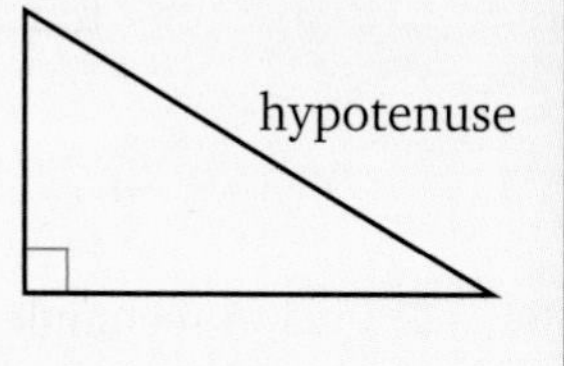

Pythagoras' theorem states that, in the right-angled triangle shown, $a^2 + b^2 = c^2$

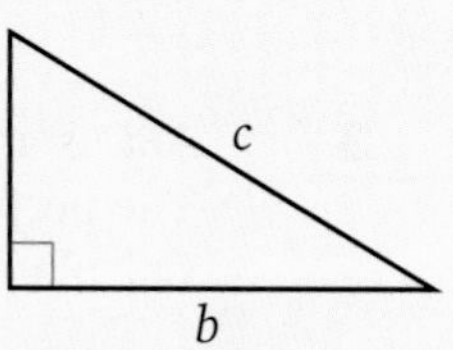

It is always the squares of the two smaller sides that add up to the square of the hypotenuse.

Right-angled triangles can be formed from other shapes, for example isosceles and equilateral triangles, and diagonals of rectangles. Wherever a right-angled triangle is formed you can use Pythagoras' theorem to find the length of the third side.

Example Pythagoras' theorem Unit 3

C

a Calculate the length of the diagonal BD in this rectangle.

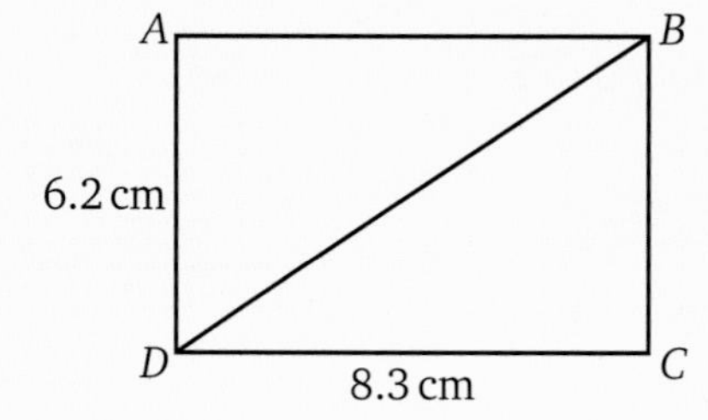

b Calculate the length of AB in the diagram.

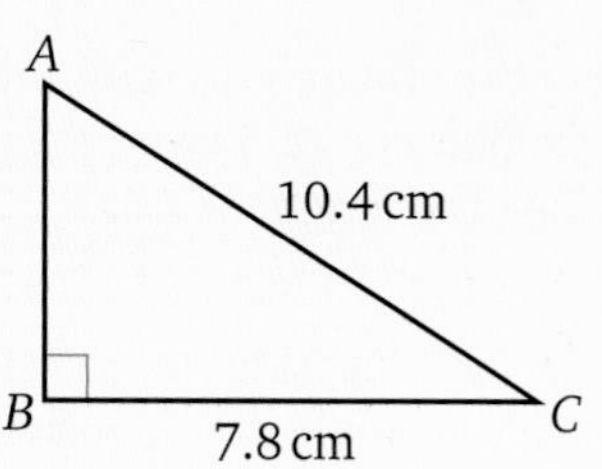

Solution

a Pythagoras' theorem states that:

$a^2 + b^2 = c^2$, where c is the hypotenuse.

In this case,

$AD^2 + AB^2 = BD^2$

$6.2^2 + 8.3^2 = BD^2$

$38.44 + 68.89 = BD^2$

$BD = \sqrt{107.33}$

$BD = 10.360019305$

$BD = 10.4$ cm (to 1 d.p.)

b Pythagoras' theorem states that:

$a^2 + b^2 = c^2$, where c is the hypotenuse.

In this case,

$AB^2 + BC^2 = AC^2$

$AB^2 + 7.8^2 = 10.4^2$

$AB^2 + 60.84 = 108.16$

$AB^2 = 47.32$

$AB = \sqrt{47.32} = 6.9$ cm (to 1 d.p.)

AQA Examiner's tip

When finding the hypotenuse you must add the squares of the other two sides, but when finding a shorter side you must subtract the squares of the other two sides.

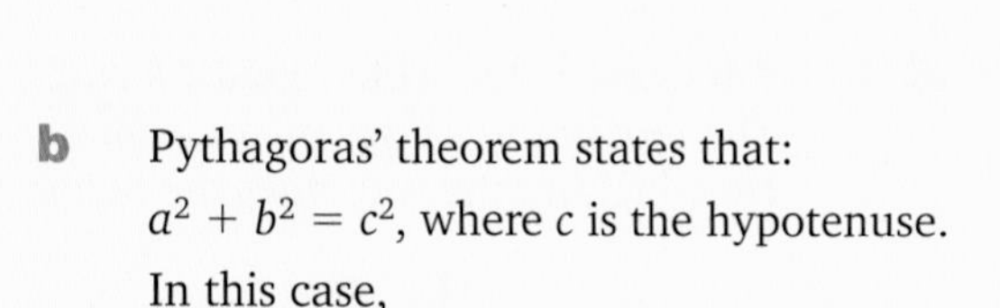

Bump up your grade

To get a Grade C make sure you understand Pythagoras' theorem and know how to apply it to solve problems.

Practise... Pythagoras' theorem Unit 3

C

C

1 Calculate the length marked x in these diagrams.

a

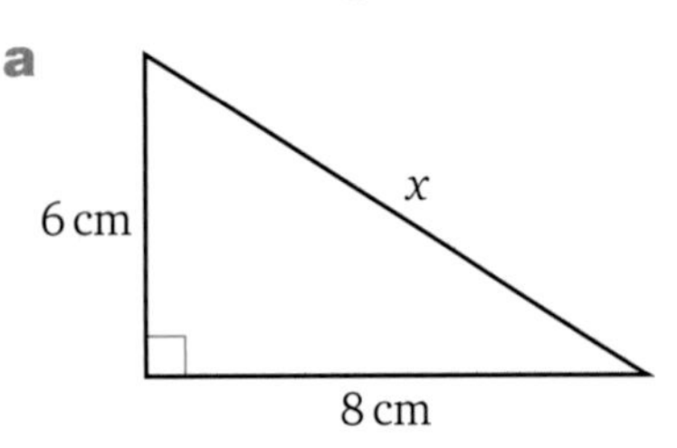

b

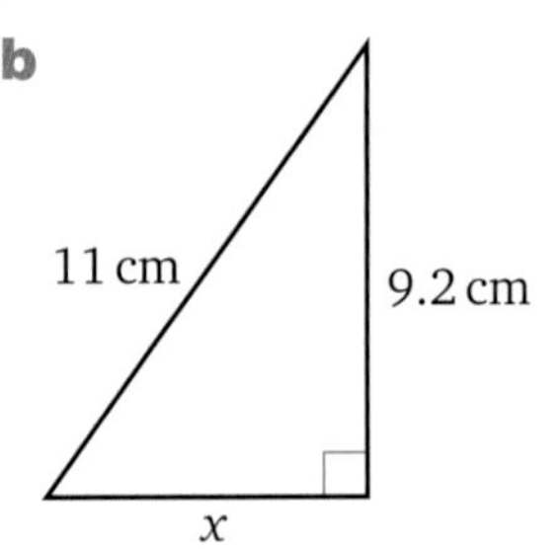

c

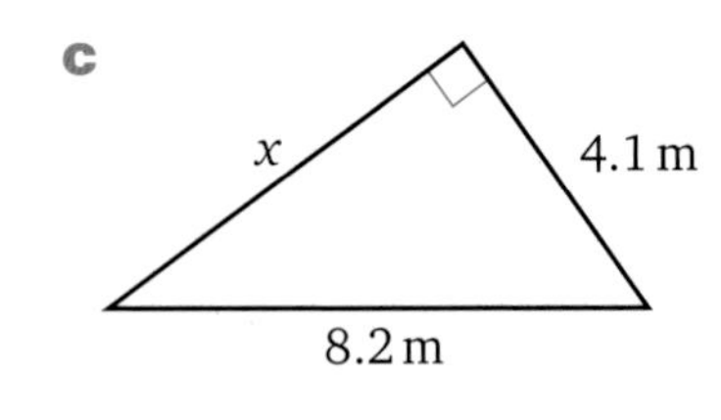

2 A rectangle measures 5.6 cm long and 3.8 cm wide.
Calculate the length of the diagonal of the rectangle.

3 ABC is an isosceles triangle with AB and AC 7.2 cm long.
The perpendicular height of the triangle is 6.9 cm.
Calculate the length of BC.

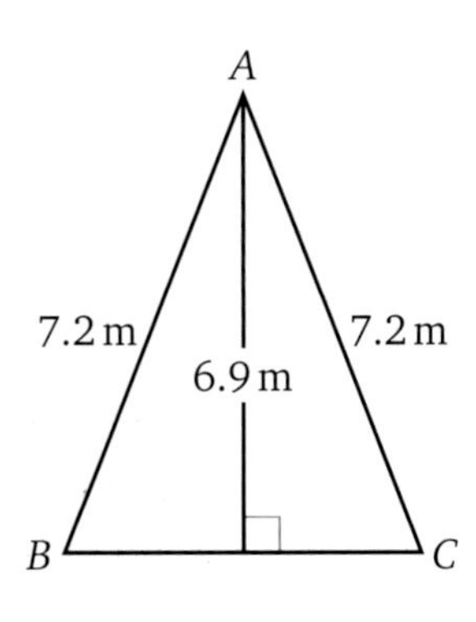

Hint

The perpendicular height splits the triangle exactly in half.

AQA Examiner's tip

Watch out for right-angled triangles 'hidden' in rectangles or isosceles triangles.

4 Draw a coordinate grid with the x-axis and y-axis from 0 to 8.
Plot the points $A(2, 7)$ and $B(6, 4)$.
Calculate the length of AB.

5 Triangle ABC has side $AB = 5$ cm, $AC = 12$ cm and $BC = 13$ cm
Show that triangle ABC is a right-angled triangle.

6 A 6 metre ladder rests against a wall.
The foot of the ladder is 2 m from the wall.
How high up the wall does the ladder reach?

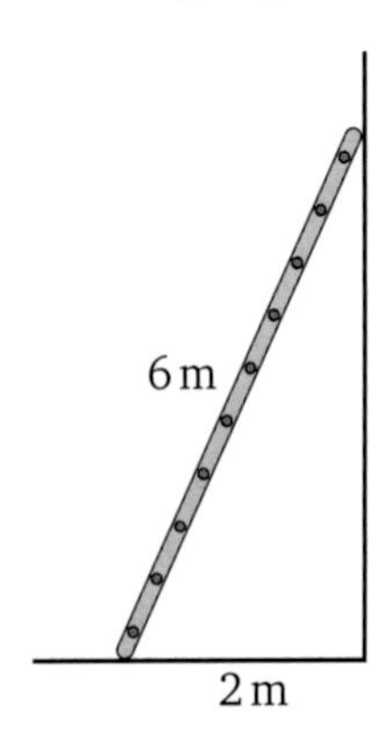

AQA Examination-style questions

1 Write down whether each statement is always true, sometimes true or never true.

a Isosceles triangles have three acute angles. *(1 mark)*

b Equilateral triangles have a right angle. *(1 mark)*

c A kite has an obtuse angle. *(1 mark)*

2 Which of these nets would make a cube? There may be more than one correct answer.

A
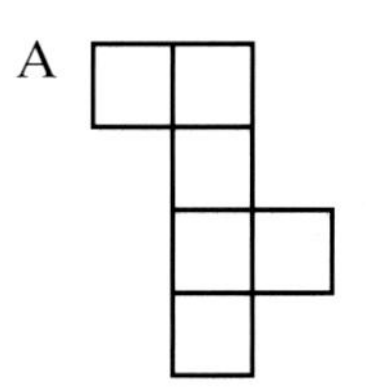

B

C
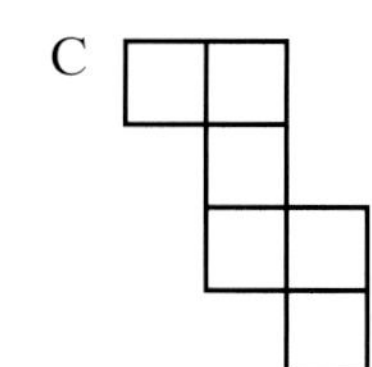

D
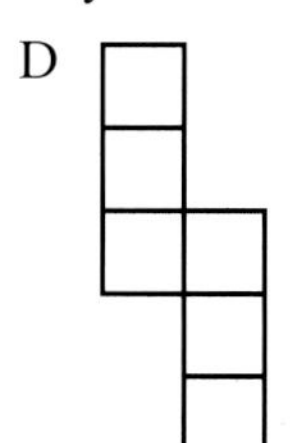

E
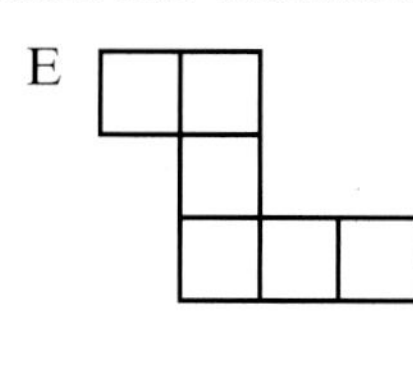

(3 marks)

3 A square and an equilateral triangle have the same perimeter.
Each side of the triangle is 8 cm long.
What is the area of the square? *(3 marks)*

4 Copy the diagram. Shade in three more squares so that the rectangle has reflection symmetry but no rotation symmetry.

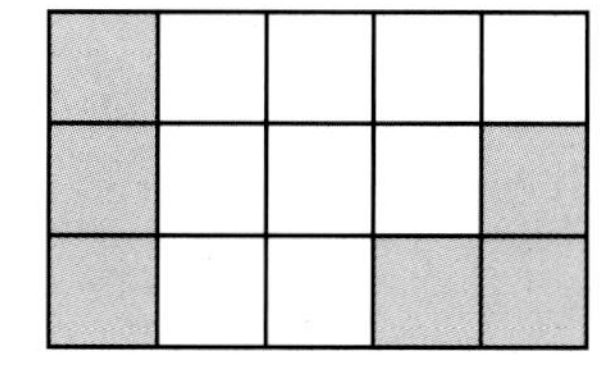

(2 marks)

5 Copy this diagram.

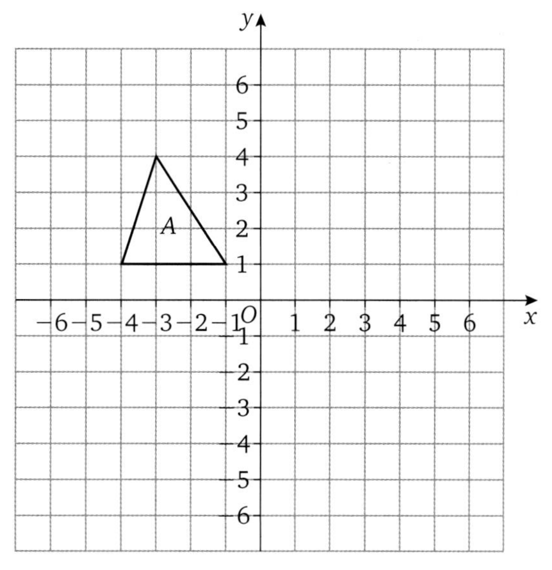

a Translate triangle A through $\begin{pmatrix} 5 \\ -6 \end{pmatrix}$. Label it B. *(1 mark)*

b Reflect triangle B in the x axis. Label it C. *(1 mark)*

6 A triangle has a base of 8 cm and an area of 24 cm^2.
Calculate the height of the triangle. *(2 marks)*

G F D

D

7 Draw the front elevation of this shape, which is made from six cubes.

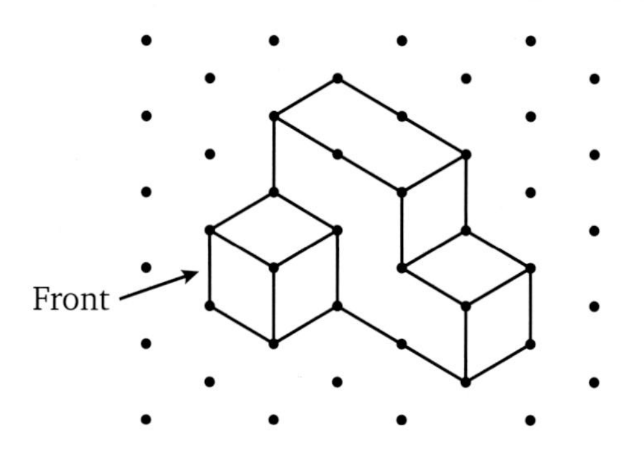

(2 marks)

8 Copy the diagram. Enlarge shape A with a scale factor of 2, with centre of enlargement $(-6, 2)$.

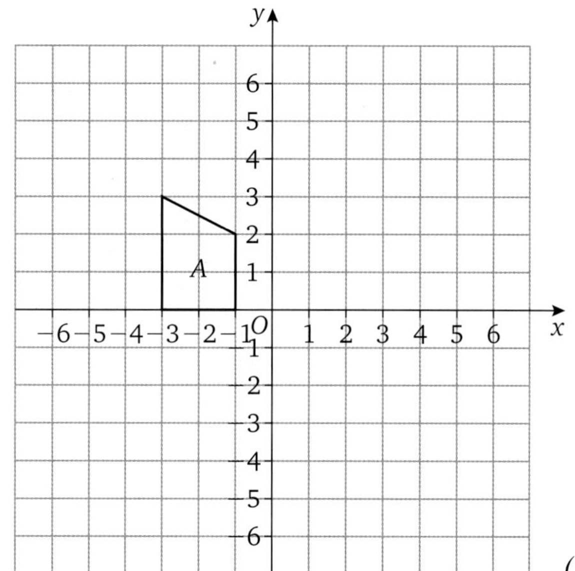

(3 marks)

C

9 **a** An isosceles triangle ABC has two sides of length 7.2 cm and a base of 5.8 cm. Construct triangle ABC. *(3 marks)*

b Calculate (do not measure) the height AD. *(3 marks)*

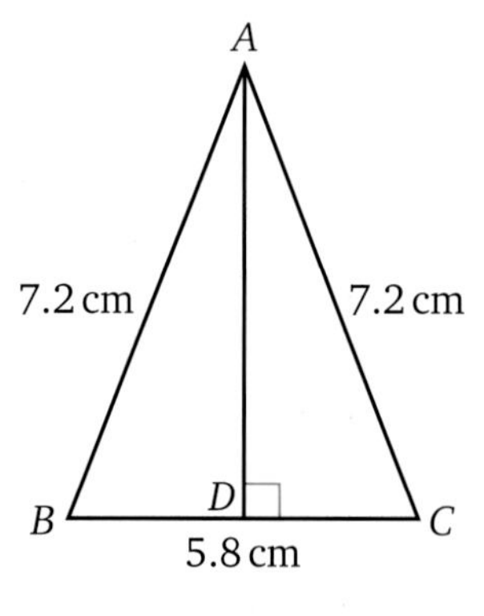

10 This tent is 2.4 m long.

The triangular cross-section has a base of 1.6 m and a height of 1.4 m.

Calculate the volume of the tent.

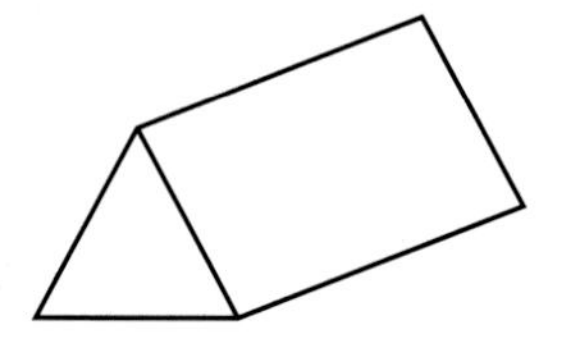

(3 marks)

11 A cake is in the shape of a cylinder.

It has a height of 6 cm and a radius of 10 cm.

Joe wants to put icing on the top of the cake and around the side.

What area of the icing does he need?

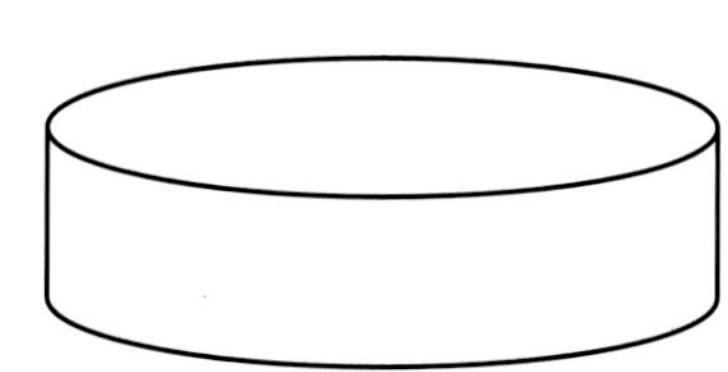

(5 marks)

Essential skills

1 Justifying answers

- Statistics and number
- Number and algebra
- Geometry and algebra

Revise... Key points

In all exams and in all units there are questions that could ask you to do things like:

- **Show** whether a statement is true or not.
- Show how a mathematical answer has been obtained.
- Make a decision and **explain why** you have made it.
- Make a decision and give a reason why you have made it.
- Show your working.

Key terms

You need to know these terms for your exam. Make sure you understand what they mean, and write your own definitions.

Explain
Explain why
Give a reason for your answer
Show
Show how you decide
Show that
Show working to justify your answer
Show working to support your answer
You **must** show your working.

Unit 1

Compare data sets
Test a hypothesis

Unit 2

Use algebra to support and construct arguments

All Units 1 2 3

Set up and solve equations

AQA Examiner's tip

To gain any marks you must follow the instructions in a question, so make sure you read the question carefully.

Example Statistics 

D

A junior school class has a spelling test on each day of the week.

The table shows Sally's scores and David's scores for the last two weeks.

	Mon	Tue	Wed	Thu	Fri	Mon	Tue	Wed	Thu	Fri
Sally's scores	8	8	7	6	8	7	10	6	7	8
David's scores	6	8	7	7	absent	absent	8	6	6	8

a Sally says that she gets higher scores than David.
Is Sally correct?
Show how you decide.

b David says his scores are more consistent.
Is David correct?
Show how you decide.

AQA Examiner's tip

In order to score any marks you must follow the instruction 'Show how you decide'.

Bump up your grade

To get a Grade C, you will need to be able to answer questions where you have to decide what you need to do.

Solution

a To work out if Sally's scores are higher you have to work out both Sally's average score and David's average score.

David has been absent for two days so you cannot use the total scores.

In this case, because you are interested in the total scores, the mean is the best average to use.

Sally's mean score $= (8 + 8 + 7 + 6 + 8 + 7 + 10 + 6 + 7 + 8) \div 10$
$= 75 \div 10$
$= 7.5$

David's mean score $= (6 + 8 + 7 + 7 + 8 + 6 + 6 + 8) \div 8$
$= 56 \div 8$
$= 7$

So Sally is correct as her mean score is 0.5 bigger than David's mean score.

b To work out whose scores are more consistent you have to compare the range of David's scores with the range of Sally's scores.

The range measures the spread of data. A smaller range means the data is less spread out and is, therefore, more consistent.

Range of scores = highest score − lowest score
David's highest score = 8
David's lowest score = 6
So the range of David's scores $= 8 - 6 = 2$
Sally's highest score = 10
Sally's lowest score = 6
So the range of Sally's scores $= 10 - 6 = 4$
So David is correct because the range of his scores is smaller than the range of Sally's scores.

AQA *Examiner's tip*
Set out your work in an organised way like this and write a conclusion.

Example Algebra Unit 2

C

a **Show that**
$2(3n + 4) + 4(n + 3)$
simplifies to
$10n + 20$

b n is a positive whole number.
Explain why $2(3n + 4) + 4(n + 3)$ is **always** a multiple of 10.

Bump up your grade
You need to know how to answer a 'Show that ...' question.
A 'Show that ...' question gives you the answer and asks you to show the **working** you would do to get the answer.

AQA *Examiner's tip*
Sometimes, the answer you have been given is used in a second part of a question.
You should always try the second part of the question even if you cannot do the first part!

Solution

a To simplify $2(3n + 4) + 4(n + 3)$ you have to expand each set of brackets first.

$2(3n + 4) = 2 \times 3n + 2 \times 4 = 6n + 8$

$4(n + 3) = 4 \times n + 4 \times 3 = 4n + 12$

The next step is to collect like terms.

Because you have been asked to 'show' try to do this as fully as possible. For example:

$2(3n + 4) + 4(n + 3) = 6n + 8 + 4n + 12 = (6n + 4n) + (8 + 12) = 10n + 20$

AQA *Examiner's tip*
In part **a** of the question you have to decide on the method you need to use. There is a clue in the word 'simplifies'.

b Two possible methods are shown below.

Method 1

When n is a positive whole number, $10n$ is $10 \times$ a positive whole number, which is **always** a multiple of 10.

Adding 20 (a multiple of 10) to a multiple of 10 must **always** give another multiple of 10.

Method 2

Start by factorising $10n + 20$

$10n + 20 = 10 \times n + 10 \times 2 = 10(n + 2)$

When n is a positive whole number $n + 2$ is also a positive whole number.

So $10(n + 2)$ is $10 \times$ a positive whole number, which is **always** a multiple of 10.

> **AQA Examiner's tip**
> Part **b** uses the answer you have been given in part **a** of the question.

Example Geometry

Amie draws this diagram.

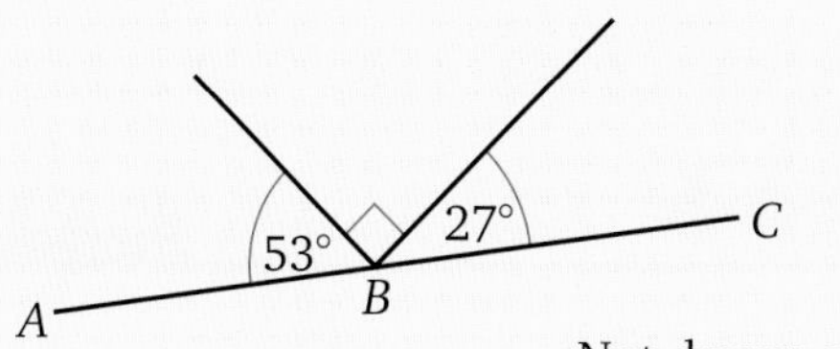

Not drawn accurately

She says that ABC is a straight line.

Is Amie correct?

Explain your answer.

> **AQA Examiner's tip**
> In order to get any marks in this question you must follow the instruction 'Explain your answer'.
> Do **not** get put off by the fact that ABC looks like a straight line. The diagram is '**not** drawn accurately'.

Solution

This question is asking if you know that angles on a straight line add up to 180°.

It is also asking if you know that the square symbol marking the middle angle represents a right angle and equals 90°.

Here is a suitable explanation:

$53 + 90 + 27 = 170$

ABC is **not** a straight line because the angles add up to 170° and, for a straight line, they need to be 180°.

Example Algebra

Show that quadrilateral $ABCD$ is a parallelogram.

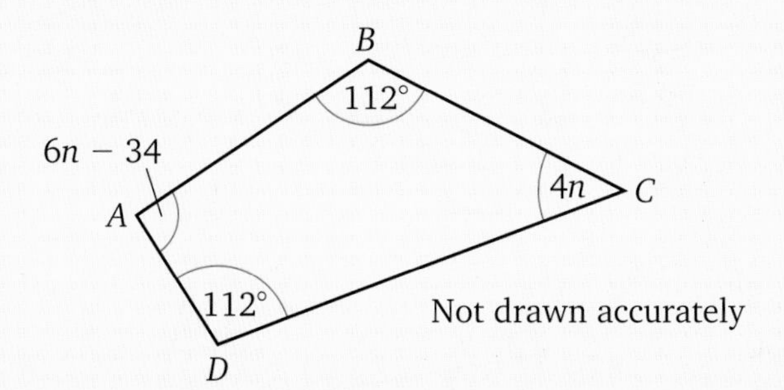

Not drawn accurately

> **AQA Examiner's tip**
> Do **not** get put off by the fact that $ABCD$ does not look like a parallelogram. It is '**not** drawn accurately'.
> You are given some information about the angles. So the question is about the angles of a quadrilateral and, in particular, the angles of a parallelogram.

Solution

The angles of a quadrilateral add up to 360°. This fact can be used to set up an equation.

$$6n - 34 + 112 + 4n + 112 = 360$$
$$10n + 190 = 360$$
$$10n + 190 - 190 = 360 - 190$$
$$10n = 170$$
$$n = 17$$

When $n = 17$, $6n - 34 = 6 \times 17 - 34 = 68$ and $4n = 4 \times 17 = 68$

So the opposite angles of *ABCD* are equal.

The opposite angles of a parallelogram are equal, so *ABCD* is a parallelogram.

You can also use the fact that the adjacent angles (112° and $4n$ or 112° and $6n - 34$) must add up to 180° to show that *ABCD* is a parallelogram.

AQA Examiner's tip

In some questions you might have to use algebra to show a geometry fact.

Look for clues about the geometry property that you need to think about before starting the question.

To get full marks, make a final statement showing which angle property of the parallelogram you are using.

Practise... Justifying answers

F

1 John thinks of a number.
He multiplies the number by 5 and then adds 12.
He gets the answer 42.
Sally thinks of a number.
She subtracts 4 and then multiplies by 9.
She gets the answer 18.
Show that John and Sally think of the same number.

2 The diagram shows two rectangles *A* and *B*.

Sam says that the perimeter of rectangle *A* is bigger than rectangle *B*.
Is Sam correct?
Show how you decide.

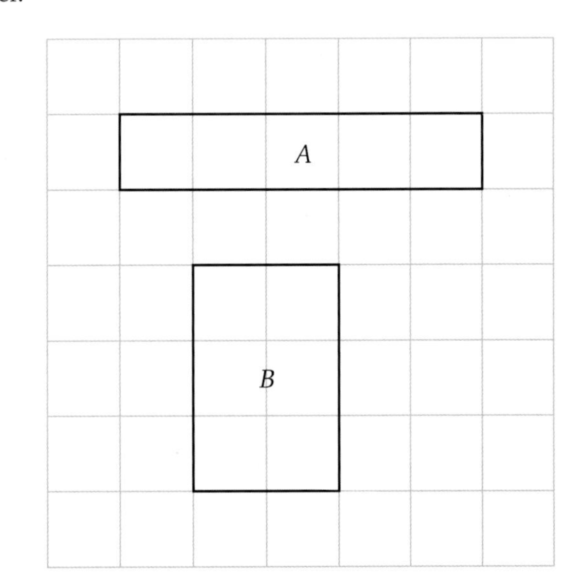

E

3 The diagram shows the sketch of a shape.

Explain why the shape is impossible to draw accurately.

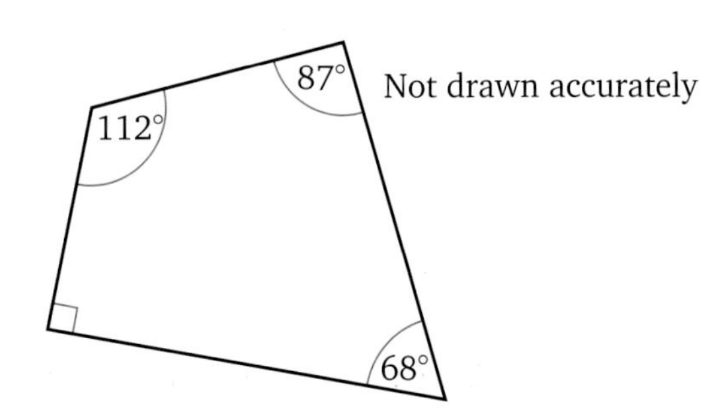

E

4 Zoe makes 200 ceramic mugs.
It costs 80 pence to make each mug.
She sells 60% of the mugs for £3 and the rest of the mugs for £2.
Zoe aims to make a profit of more than £350.
Does she do this?
You **must show working to support your answer**.

5 This is an addition pyramid

15
7 8

The number in a box is the sum of the numbers in the boxes immediately underneath.
These algebraic addition pyramids work the same way.

Pyramid **A**

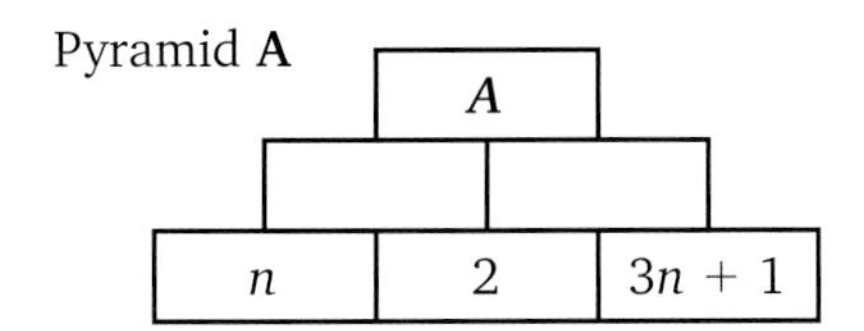

Pyramid **B**

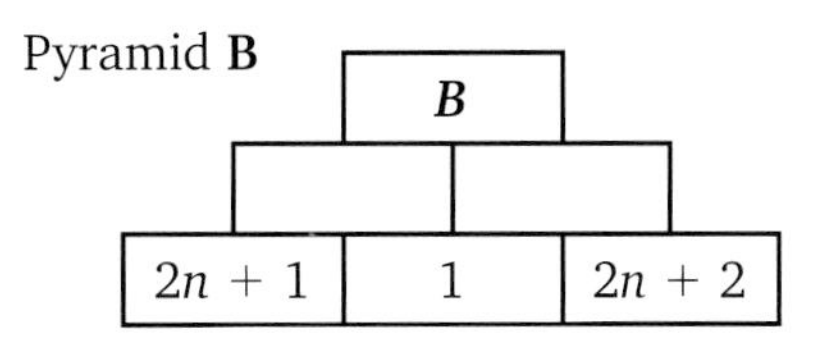

Show that the expression in box ***A*** equals the expression in box ***B***.

D

6 **a** Chandi has a dice with 1 red face, 2 white faces and 3 blue faces.
She throws the dice 600 times.
The colours she gets are shown in the table.

Red	White	Blue
98	196	306

Chandi says the dice is fair.
Is she right?
Show how you decide.

b Emma has a dice with 2 red faces and 4 blue faces.
She throws the dice 10 times and gets 1 blue.
Emma says the dice is **not** fair.
Explain why she could be wrong.

7 A Year 8 class have a mental maths test every Monday.
The scores for the first three weeks of term for the boys in the class are shown in the table.

Boy's mental maths test scores (out of 20)

Name	Week 1	Week 2	Week 3
Sam	15	15	15
Tom	13	abs	18
Bill	17	12	abs
Eric	17	17	19
Rav	10	11	14
David	11	20	18
Avi	12	14	abs
James	13	15	17
Paul	8	9	11
Mike	16	16	15
Terry	15	17	14

Compare the boys' scores in these three weeks.

D

8 These two number machines **always** have the same input.

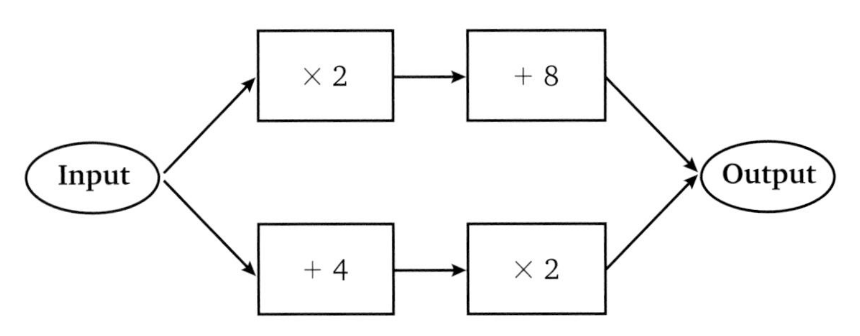

Show that they **always** have the same output.

9 X is an odd number.
Y is an even number.

a Explain why $2X + Y$ is **always** an even number.

b Tim says: '$X + 2Y - 3$ **cannot** be a prime number'.
Is Tim correct?
Explain how you know.

C

10 $4(2x + 3) - 6(x + 1)$ simplifies to $a(x + b)$

a Show that $a = 2$ and $b = 3$

b Explain why $4(2x + 3) - 6(x + 1)$ is even when x is a positive whole number.

11 The diagram shows a rectangle and a square.
The length of the side of the square is $2x - 3y$
The length of one side of the rectangle is $x + 2y$

$2x - 3y$

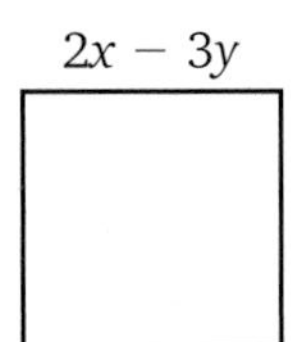

$x + 2y$

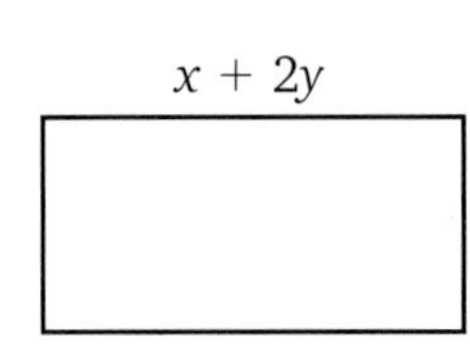

Not drawn accurately

The rectangle and the square have the same perimeter.
Show that the length of the other side of the rectangle is $3x - 8y$.

12 The diagram shows triangle ABC.
D is a point on AC.
Angle $BAD = 55°$, angle $BCD = 30°$ and angle $DBC = 25°$

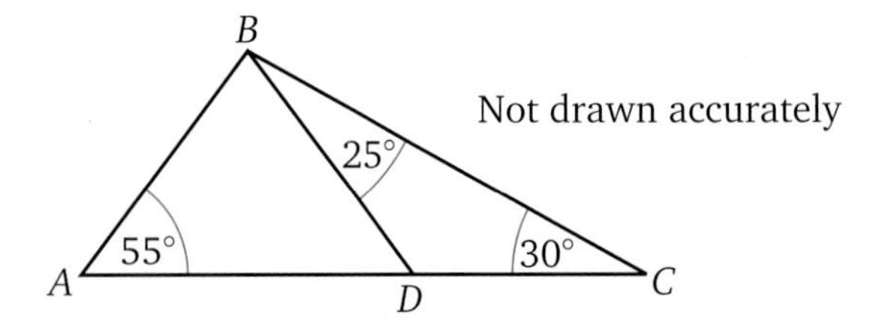

Show that the length of AB and the length of BD are equal.

Essential skills

2 Problem solving

- Statistics and number
- Algebra and Number
- Algebra and geometry

Revise... Key points

Questions in your maths exams are not always about a particular piece of mathematics that you have learnt.

In some questions you have to use the maths you know to solve a problem about mathematics or real life.

In these questions you are **not** told the mathematical method that you have to use. You have to choose it yourself.

You have to stop, think and try to puzzle out what you need to do.

Here are some of the things you might try.

- Working systematically. This means being methodical and organised. It also means trying to break up a problem into its separate parts and then doing one part at a time.
- Working backwards. This means trying out familiar inverse operations using a value given in the problem to see if it helps.
- Finding examples that fit. This is useful whenever an answer has to fit more than one condition. Start by trying a value that fits one of the conditions and see if it fits any of the others. Then change the value systematically until it fits all of the conditions.
- Finding a relationship. Try out connections between values given in the question and see if they help.
- Choosing the maths. This means choosing the maths that you could use to help you solve the problem, then trying it to see if it does.

Key terms

These terms will be used in your exam. Make sure you understand what they mean, and write your own definitions.

Explain
Explain why
Give a reason for your answer
Show
Show how you decide
Show that
Show working to justify your answer
Show working to support your answer
You **must** show your working

Unit 1

Compare data sets
Test a hypothesis

Unit 2

Use algebra to support and construct arguments

Set up and solve equations

Bump up your grade

You will need to be able to answer problem-solving questions to get a Grade C.

AQA Examiner's tip

Sometimes problem-solving questions have more words than other questions. So it is important to read the question carefully, one sentence at a time, and try to pick out the important pieces of information.

Do this a few times until you are sure what the question is asking you to do.

When you have finished your answer, read the question again to make sure you have done all that it has asked.

Example Statistics Unit 1

E

Pie chart A shows the breakfast cereal eaten by 80 students on Monday morning.

Pie chart B shows which of these students choose Crispies with sugar.

Pie chart C shows which of the students choose Bix with sugar.

A

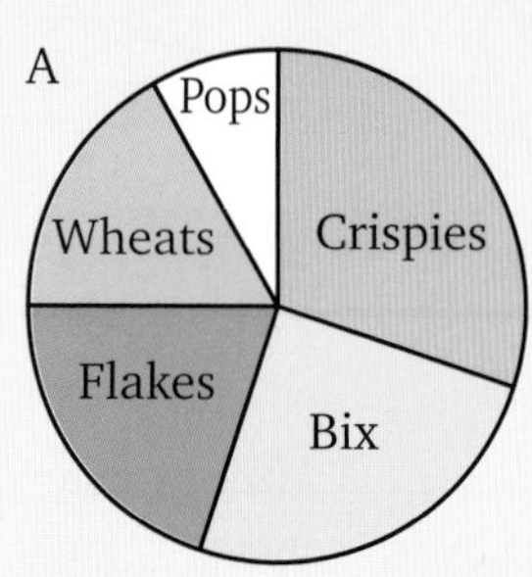

B

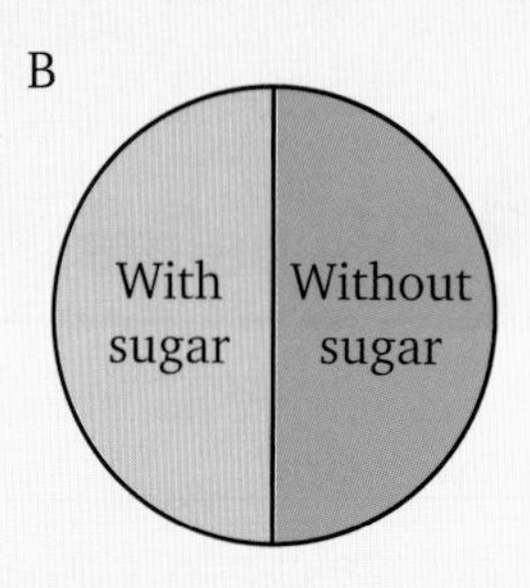

C

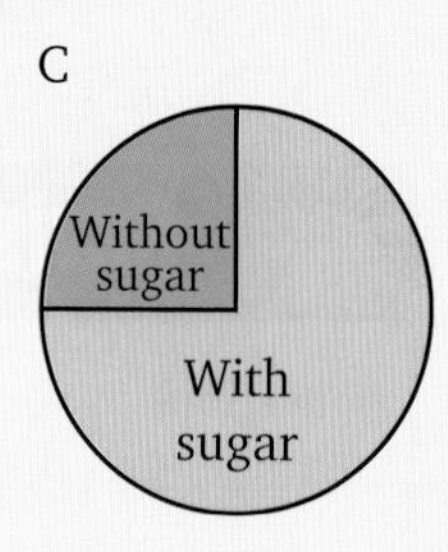

Compare the number of students who eat Crispies with sugar with the number of students who eat Bix with sugar.

Solution

The angle for Crispies on pie chart A is 108°.

So the number of students eating Crispies $= \frac{108}{360} \times 80 = 24$

Pie chart B shows that $\frac{1}{2}$ the students who eat Crispies do so with sugar.

So 12 students eat Crispies with sugar.

The angle for Bix on pie chart A is 90°.

So the number of students eating Bix $= \frac{90}{360} \times 80 = 20$

Pie chart B shows that $\frac{3}{4}$ of the students who eat Bix do so with sugar.

$\frac{3}{4}$ of $20 = 15$

So 15 students eat Bix with sugar.

Three more students eats Bix with sugar than eats Crispies with sugar.

AQA Examiner's tip

Break the task down into small steps.

The instruction 'compare' means you must complete the question by stating the differences and/or similarities.

Example Number 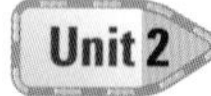

F

You can make four-digit numbers using these cards.

5	0	8	1

The number shown is 5081.

Use the cards to make the smallest multiple of 5 greater than 1000.

AQA Examiner's tip

This question asks you to write down a number that has to meet three conditions.

The best way to try and find this number is to work **systematically** and look at one condition at a time.

Solution

For the number to be greater than 1000:

The thousands digit must be 1, 5 or 8

For the number to be the smallest possible, the thousands digit must be the smallest possible:

So the thousands digit is 1

For the number to be a multiple of 5:

The units digit must be 0 or 5 A multiple of 5 is a number in the 5 times table.

For the number to be the smallest possible, the zero must have the largest possible place value:

So the hundreds digit **must** be 0 and the units digit must be 5

This means that the tens digit **must** be 8

So the smallest multiple of 5 greater than 1000 that can be made with the cards is 1085.

Example Algebra Unit 2

D

There is a whole number on the back of each of the cards A, B and C.

The number on card B is double the number on card A.

The number on card C is 10 less than the number on card B.

The sum of the numbers on cards A, B and C is 210.

Work out the number on card C.

A	B	C

AQA Examiner's tip

You can try values to see if they work or you can **set up and solve an equation**. Setting up and solving an equation is the most efficient method and is likely to take less time.

Solution

Let the number on card A $= x$

So the number on card B $= 2x$

and the number on card C $= 2x - 10$

The total of the numbers on all three cards is $x + 2x + 2x - 10$

This simplifies to $5x - 10$

The total of the numbers on all three cards is **given** as 210, so the equation is

$5x - 10 = 210$

So $5x = 220$ adding 10 to both sides

and $x = 44$ dividing both sides by 5

The number on card C $= 2 \times 44 - 10 = 78$

AQA Examiner's tip

Always check the answers to problems like this.

Card A = 44, Card B = $2 \times 44 = 88$, Card C = $88 - 10 = 78$

$44 + 88 + 78 = 210$

Practise... Problem solving Unit 1

F E D C

F

1 Altogether Sam saves £305 in the first six months of 2010.
The bar chart shows the amount of money that he saves in each of the first five months of 2010.

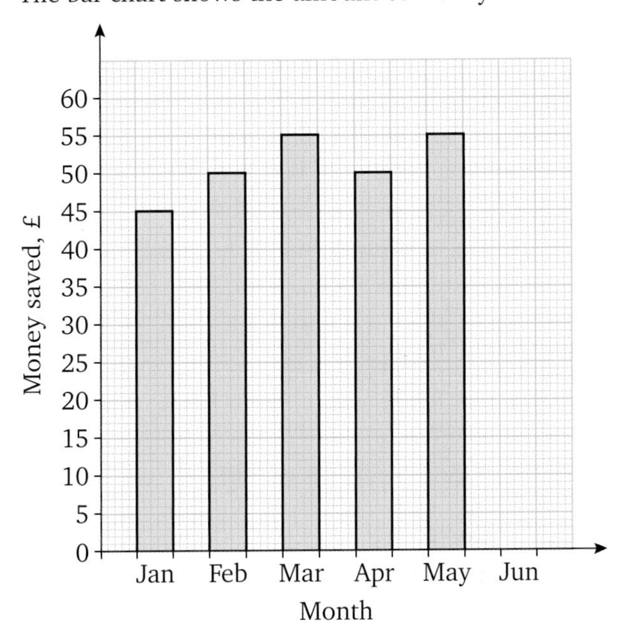

Work out the mode of the amounts that Sam saves each month in the first 6 months of 2010.
Show how you decide.

E  **2** At a gym a notice on the rowing machine says: '*On this machine you use 6 calories a minute*'.
On the cross trainer it says: '*On this machine you use 9 calories a minute*'.
Tom trains for an equal amount of time on the rowing machine and the cross trainer.
He uses exactly 300 calories.
Work out the total amount of time that he trains.

D   **3** Charlotte has four different coins in her pocket.
She picks three of the coins at random.
It is **not** possible for the total amount to be more than 72 pence.
Write a list of all the possible total amounts of money Charlotte can take from her pocket.
You must **show** your working.

4 Rectangles **A** and **B** have equal perimeters.

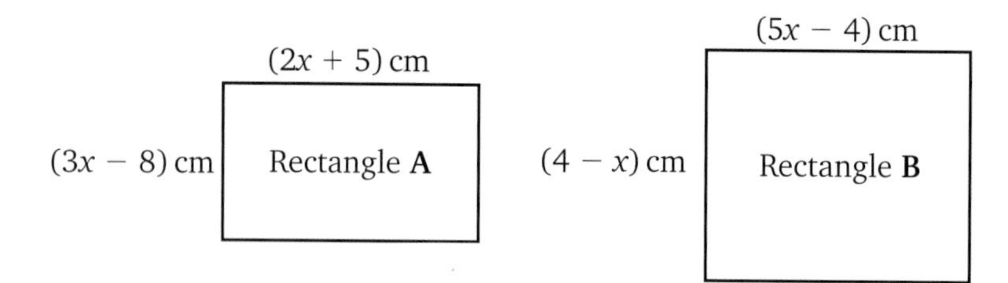

Which rectangle has the greatest area?
Show how you decide.

C  **5** Bill thinks of a number.
He subtracts 3 from his number and then multiplies by 4.
Sally thinks of the same number as Bill.
She multiplies the number by 2 and then adds 15.
Bill and Sally get the same answer.
What number do they both think of?

 **6** It takes a gardener $2\frac{1}{2}$ minutes to tidy up one metre of the edge of some flower beds.
How long does it take the gardener to tidy up a circular flower bed of diameter 3.2 metres?
Give your answer to the nearest minute.

AQA Examination-style questions

AQA Examination-style questions

1 Lucy buys two chocolate bars costing 65p each.
She pays with a £5 note and receives the correct change.
She is given **exactly** five coins.
What could these coins be? *(3 marks)*

G

2 A number is written on the other side of each of these cards.

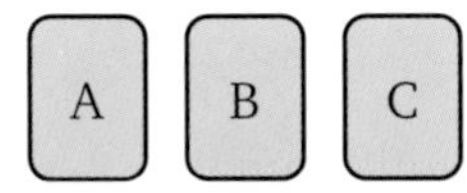

There is a whole number on the other side of each card.
The sum of the numbers on all three cards is 15.
The sum of the number on card **A** and the number on card **B** is 7.
The difference between the number on card **A** and the number on card **C** is 5.
Work out the number on each of the cards. *(3 marks)*

F

3 Tom makes a cuboid using fifteen 1 cm cubes.
Jerry makes a different cuboid using fifteen 1 cm cubes.
Tom's cuboid has a smaller surface area than Jerry's cuboid.
What are the surface areas of Tom and Jerry's cuboids?
You **must** show your working. *(4 marks)*

D

4 Donna tries to solve puzzles at different levels of difficulty, from level 1 (easy) to level 10 (hard).
The table shows the times she takes to solve **some** of the puzzles.

Level of difficulty	1	2	3	4	5	6	7	8	9	10
Time (minutes)	6	11	13	18	25	24		38	45	53

a Donna is interested in the relationship between the level of difficulty and the time she takes to solve the puzzles. On a suitable grid draw a diagram to show the relationship.

b Estimate the time it takes Donna to solve a level 7 puzzle. *(3 marks)*

C

Number

1 Answers

1 Types of numbers

1 a Twenty-three thousand, six hundred and seventy-two.
b i 70 **ii** 3000
c 23 672

2 a i 5007 **ii** 7500 **iii** 5700
b i 75 000 **ii** 750

3 a i 727 **ii** 616 **iii** 962 **iv** 107

4 a Friday **b** Wednesday **c** 5 °C

5 a $8 - (4 + 1) = 3$
b $(5 + 2 - 4) \times 2 = 6$
c $(8 + 6) \div (3 - 1) = 7$
d $(9 - 6 \times 4) \div 3 = -5$

6 Example showing a negative number subtracting another negative number to give a negative answer.

7 a False. Although 43 is prime, 27 is not prime as it has more than 2 factors (1, 3, 9, 27).
b False. $27 + 43 = 70$ which is even.
c False. 43 does not divide exactly by 3.
d True. $43 - 27 = 16$ which does divide exactly by 4.

8 a 56 **b** 9, 27 **c** 9 **d** 19 **e** 24

9 1, 2, 3, 6

10 a i 1, 2, 7, 14 **ii** 1, 5, 7, 35
b 7
c 70

11 a $54 = 2 \times 3^3$ **b** 18

12 If $p = 2$ (prime), then $3p = 6$ which is even, not odd.

13 a Must be odd, since $2p$ is even.
b If p is odd, so is p^2, so $p^2 + 1$ must be even.

14 Classic is cheaper for the whole week.
(4 × £4 more during the week, but 3 × £6 less at the weekend means that overall it is £2 cheaper.)

15

Balance
−£45
−£81
£39
−£36
£946

16 The LCM of 10, 6 and 4 is 60. Kate must buy 6 packs of pencils, 10 packs of notepads and 15 packs of pencil sharpeners.

2 Fractions, decimals and rounding

1 **a** $\frac{2}{3}$ **b** $\frac{1}{3}$

2 **a** $\frac{1}{2}$ **b** $\frac{4}{5}$ **c** $\frac{3}{5}$ **d** $\frac{3}{10}$ **e** $\frac{3}{4}$

3 **a** 73 **b** 80 km **c** £72 **d** 535 kg

4 $\frac{7}{12}$, $\frac{2}{3}$, $\frac{3}{4}$, $\frac{7}{9}$, $\frac{5}{6}$

5 0.667, 0.67, 0.676, 0.7, 0.759, 0.76

6 **a** 83 000 **b** 82 800 **c** 82 800

7 **a** **i** 26.32 **ii** 3.22

8 **a** $1\frac{2}{21}$ **b** $\frac{1}{5}$

9 **a** 200 **b** Over-estimate as both numbers were rounded upwards.

10 **a** 3.847926267 **b** **i** 3.8 **ii** 4

11 **a** 0.7 **b** 0.8 **c** 0.75 **d** 0.65 **e** $0.\dot{6}$

12 **a** $\frac{9}{14}$ **b** $1\frac{1}{2}$

13 **a** $\frac{3}{5}$ **b** $\frac{1}{4}$ **c** $\frac{17}{20}$ **d** $\frac{14}{25}$ **e** $\frac{3}{8}$

14 **a** $\frac{2}{5}$ **b** $\frac{3}{5}$

15 **a** **i** 0.72 **ii** 12.6

16 **a** $7\frac{1}{10}$ **b** $3\frac{5}{6}$ **c** $8\frac{4}{5}$ **d** $\frac{3}{4}$

17 **a**, **e** and **g**.

18 Yes $1\frac{1}{3} \times \frac{2}{3} = \frac{4}{3} \times \frac{2}{3} = \frac{8}{9}$ is less than 1.

19 **a** 4 **b** £3.20

20 £214.40

21 5 tins

22 **a** $\frac{11}{20}$ **b** $0.18\dot{3}$

23 **a** 29 pence **b** 20p, 5p, 2p, 2p

3 Percentages

1 Shape A **a** 25% **b** 75%
Shape B **a** 50% **b** 50%
Shape C **a** 60% **b** 40%
Shape D **a** 40% **b** 60%

2

Decimal	Fraction	Percentage
0.3	$\frac{3}{10}$	30%
0.85	$\frac{17}{20}$	85%
0.375	$\frac{3}{8}$	37.5%
0.75	$\frac{3}{4}$	75%
0.4	$\frac{2}{5}$	40%
$0.\dot{3}$	$\frac{1}{3}$	$33\frac{1}{3}$%
0.07	$\frac{7}{100}$	7%
0.28	$\frac{7}{25}$	28%
0.025	$\frac{1}{40}$	2.5%
2.5	$2\frac{1}{2}$	250%
1.25	$1\frac{1}{4}$	125%
4.2	$4\frac{1}{5}$	420%

3 45%, 0.54, $\frac{4}{5}$

4 **a** £140 **b** 840 m **c** 9 kg **d** 14 litres

5 **a** £10.20 **b** 153 km **c** £1.41 **d** £19.20

6 **a** **i** £4.80 **ii** £3.20

7 **a** **i** £143.10 **ii** £126.90

8 Rob (Rob's serves 40% aces, Simon serves 35% aces).

9

Vitamin	Bowl contains	RDA	%
Iron	4.5 mg	14.0 mg	32
Thiamine	0.5 mg	1.4 mg	36
Riboflavin	0.7 mg	1.6 mg	44
Niacin	5.9 mg	17.9 mg	33

10 **a** 20% **b** 25%

11 Jack's (4%, more than Jill's 3.5%).

12 **c** has the most (80% whereas **a** has 75% and **b** has 70%).

13 £18 480

14 **a** £7030.40 **b** £530.40

15 £9

16 **a** £198 **b** £30

17 Ruth (Sheila's height is only 84% of Toby's).

4 Ratio and proportion

1 **a** 3 : 2 **b** 5 : 8 **c** 1 : 4 **d** 3 : 4 : 7

2 **a** £40, £60
b £1620, £3780
c 7 miles, 14 miles, 21 miles
d 30kg, 150 kg, 180 kg

3 **a** 28 **b** 12 **c** 18 **d** 25

4 **a** $\frac{5}{9}$ **b** $\frac{4}{9}$ **c** 15 girls, 12 boys

5 65%

6 £2.40

7 **a** 50 people **b** 96°

8 Will gets £16, Kate gets £20 and Neil gets £28.

9 1.5 m^3 of sand, 0.5 m^3 of cement and 3 m^3 of aggregate.

10 8.5 km

11 **a** School A 1 : 15.6 School B 1 : 14.8 School C 1 : 14.2 School D 1 : 15.0
b 1 : 14.2 (School C) because it is likely to have smaller class sizes.

12 **a** Large bag gives the best value for money (small 0.18p/g, medium 0.161p/g, large 0.1595p/g).
b She may not have enough room or money for the large bag.

13 5 pence

14 Adel gets 12%, Barry gets 18%, Cathy gets 30%.

5 Indices

1 **a** 64 **b** 81 **c** −81 **d** 81 **e** 11 **f** 13 **g** 900 **h** 30 **i** 0.36 **j** 0.6

2 **a** 130 **b** 98

3 **a** 1000 **b** 8 **c** −8 **d** −8 **e** 4 **f** 2 **g** 3 **h** −10

4 **a** 1 **b** 10 **c** 0 **d** 10 010

5 No, with an example such as 64 (= 8^2 and 4^3) or 729 (= 27^2 and 9^3).

6 **a** 92 **b** $\sqrt{36}$ **c** $\sqrt[3]{-27}$

7 **a** 48 **b** 10 **c** 120 **d** −1 **e** 9 **f** −2

8 **a** 10 000 **b** 10 000 **c** 32 **d** −32 **e** 625 **f** 1024

9 **a** 3^9 **b** 7^4 **c** 6^6 **d** p^{10} **e** q **f** r^{15} **g** 5^2 **h** 4^{21} **i** 8^{24} **j** 9^4 **k** t **l** n^6

10 **a** 144 **b** 40 **c** 9 **d** 2 **e** 1250

11 **a** Chloe has multiplied the indices instead of adding them. Dave has multiplied the base numbers.
b 3^7 or 2187.

12 £360 – £400

13 **a** The unit digit will always come from $1 \times 1 = 1$
b 0, 5 and 6.

14 **a** **i** 4 **ii** 9 **iii** 16
b $100^2 = 10\,000$

15 Any terms with indices that equal 81 (e.g. 9^2, 3^4, $3^3 \times 3^1$).

AQA Examination-style questions

1 **a** 87 521
b Any three-digit number ending in 2 or 8.
c Any three-digit number starting with 5.
d 72 and 18 or 78 and 12.
e 52 – 27 or 82 – 57

2 **a** 80 pence or £0.80.
b 3 cones, 1 choc ice + 1 tub, 1 choc ice + 2 lollies
c 48

3 36 is the only even number, 49 is not a multiple of 3, 57 is not a square number (or other reasons).

4 **a** $\frac{2}{9}$
b $\frac{5}{8}$ since eighths are smaller than sevenths or $\frac{5}{7} = \frac{40}{56}$ and $\frac{5}{8} = \frac{35}{56}$ or $\frac{5}{7} = 0.714...$ and $\frac{5}{8} = 0.625$
c 166.7

5 **a** $\frac{8}{12}$ **b** **i** $\frac{3}{4}$ **ii** $\frac{1}{2}$ **iii** $\frac{5}{9}$

6 Option A (£17.20 for extra calls and texts, whereas Option B costs £18 more for extra calls).

7 Yes. She will have £16.80 left and the handbag costs £16.66.

8 a $(8 + 6 - 4) \div 2 = 5$ **b** $(9 - 2) \times (4 + 1) = 35$

9 a 4°C **b** $\frac{4}{7}$

10 a £552 **b** 3.2%

11 £1400

12 a Shop Q (£110) as it costs £130 at Shop P and £112 at Shop R.
b Shop P needs a smaller payment at the start/gives the longest time to pay.
Shop Q gives next day delivery. Shop R is the cheapest when you include delivery.

13 2 + any other prime number.

14 a i 0.16 **ii** 8 **b i** 18 000 **ii** 8 (or 10)

15 a The factors of 28 are 1, 2, 4, 7, 14, 28. The factors of 42 are 1, 2, 3, 6, 7, 14, 21, 42.
b 14

16 a ± 68 **b** 4.913 **c i** 12.88194444 **ii** 12.9 **d** $\frac{1}{64}$ or 0.015625

17 a 30
b ii The square of any odd number is odd, and an odd number – another odd number is always even.

18 9 tins

19 a $2^3 \times 3^2$ **b i** $2^4 \times 3^2$ **ii** $2^4 \times 3^2 \times 5$

20 12 times

21 Offer A is better (0.252p/g is cheaper than 0.277p/g).

22 20%

23 £304

24 60 g

25 a 3^{10} **b** 9

26 a 200 **b** $x = 5, y = 3$

27 a $\sqrt[3]{64} = 4$ and $\sqrt[3]{125} = 5$ and $\sqrt[3]{100}$ is between these.
b $\sqrt[3]{-1000}$ (-10 is below -5.)

28 a Any calculation involving two whole numbers with an answer of 3.5, e.g. $7 \div 2$
b Any calculation involving three whole numbers with an answer of 3.5, e.g. $(3 + 4) \div 2$, $3 + 1 \div 2$
c 0.35

Statistics

2 Answers

1 Collecting data

1 a Tally chart with the following frequencies: Red 8, Green 5, Blue 4, Orange 3.
b Red.
c Discrete, because it can only take fixed values (e.g. red...).
Qualitative as it is non-numerical data.

2 a

Money	Tally	Frequency
$£0 \leqslant x < £5$	||||	4
$£5 \leqslant x < £10$	~~||||~~ |||	8
$£10 \leqslant x < £15$	||||	4
$£15 \leqslant x < £20$	|	1
£20+	|	1

b $\frac{2}{3}$
c Discrete data.

3 a 25 **b** 29 **c** 28% **d** $\frac{11}{29}$

4 No, it is not true. Fifty-four students get a higher grade in English than maths, whereas 98 obtain a better mark in maths than English.

5 Method 1: Unlikely to get a representative set of views (e.g. unlikely to get people who work or young people).
Method 2: Not totally random as not everyone is in the phone book. Some people will therefore have no chance of being selected.
Method 3: May not be representative if survey only carried out in one part of town.

6 a Leading question/biased towards Pricewise.
b There is no 'Disagree' box.
c Asking only customers in Pricewise, so biased.

7 C, A, B, D

2 Statistical measures

1 a 7 **b** All the cards are different. **c** Sarah.

2 a 3.4... **b** 3 **c** 7

3 One possible set of values is 1, 3, 6, 7, 13. (The mean must be 6 and the range must be twice the mean.)

4 a Mean must lie between 3 and 6.
b 5.01 (to 2 d.p.).

5 a 1 letter **b** 2 letters **c** 2 letters **d** $7 - 0 = 7$ letters

6 Mean height of basketball players is 1.94 m and range is 0.14 m.
Mean height of football players is 1.81 m and range is 0.31 m.
The basketball players on average are taller. There is a greater range of heights in the football team.

7 a 1.5 people per car (rounded to 1 d.p.).
b 1.8 people per car (rounded to 1 d.p.).
c It may have had some affect as there are more people going into the city in fewer cars.

8 a $40 < x \leqslant 100$ **b** $20 < x \leqslant 40$ **c** 61 texts.

3 Representing data

1 a 18 customers.

b Five and a quarter sweet symbols drawn.

c 27 people.

2 Both yes and no are permitted with reasons.
The fire engine is at least twice as busy on Saturday as on any other weekday. Sunday is at least twice as busy as every other day except Wednesday. However, call-outs midweek total 15, and the total of weekend call-outs is 22. Using this information it is not twice as busy at the weekend.

3 a Bar drawn to height of 2. **b** 33 goals.

4 Pie chart labelled with the following: Walk 84°, Car 90°, Bus 147°, Other 39°.

5 a 13 **b** 2 **c** 17 seconds. **d** 34 seconds.

6 a

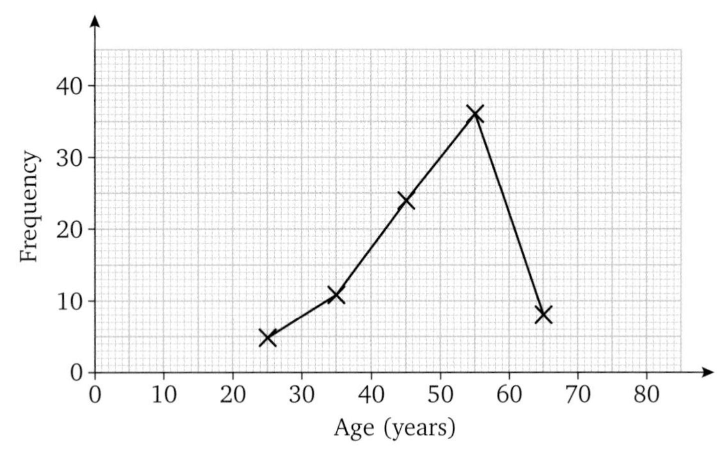

b The first class is more popular.

c On average younger people attend the second class. You can tell this as the mode is lower down the graph. The first class has a larger range as the diagram is more spread out.

7 a £2.75 **b** £2.10 **c** 2005 to 2006, increase is 85p. **d** 130% increase.

8 a

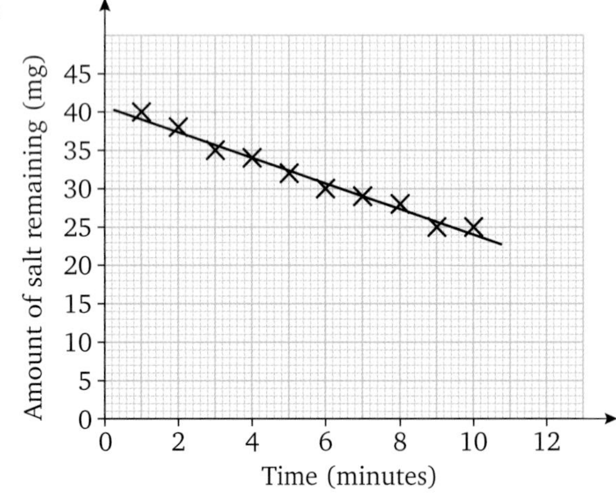

b 19–21 milligrams of substance left at 14 minutes. Not reliable as trend may not continue.

4 Probability

1 $\frac{6}{8}$ or $\frac{3}{4}$ or 0.75.

2 a

Dice score

+	1	2	3	4	5	6
3	4	5	6	7	8	9
4	5	6	7	8	9	10
5	6	7	8	9	10	11

Spinner score

b $\frac{12}{18}$ or $\frac{2}{3}$ **c** $\frac{9}{18}$ or $\frac{1}{2}$

3 Any even number above 5.

4 a 0.7 **b** 0.1 **c** 0.7 **d** 36

5 27 games.

6 a 5

b

Score	1	2	3	4	5	6
Relative frequency	$\frac{3}{30}$	$\frac{5}{30}$	$\frac{14}{30}$	$\frac{3}{30}$	$\frac{3}{30}$	$\frac{2}{30}$

c No, the results suggest the dice may be biased as number 3 has come up nearly half the time. (Allow the answer Yes – need to repeat the experiment more times.)

7 a $\frac{3}{10}$ or 0.3 **b** $\frac{11}{40}$ or 0.275

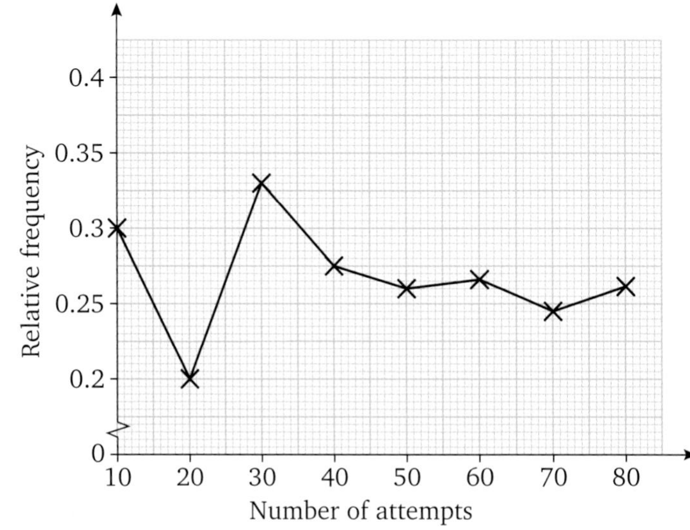

d 0.2625 (allow between 0.25 and 0.27).

8 One possible answer would be 32 red, 48 blue and 40 green counters.

5 AQA Examination-style questions

1 a Tally chart drawn with the following tallies and frequencies

Meal	Tally	Frequency
Fish and chips	||	2
Lasagne	||||	4
Roast chicken	||	2
Beef burger	~~||||~~ |	6

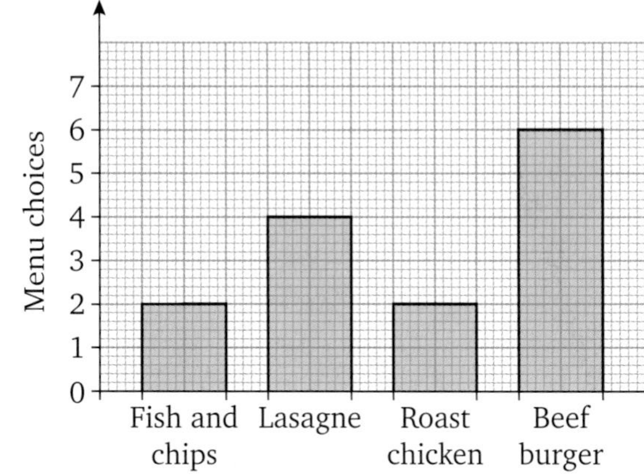

2 1 red, 1 green and 8 blue OR 2 red, 2 green, 6 blue OR 3 red, 3 green and 4 blue.

3 a 7.6 **b** 22

4 Total DVDs = 159 Total audio books = 133

5 a Saturday **b** Wednesday **c** 8.3% **d**

Sat 108° Wed 60° Thur 90° Fri 102°

6 a $\frac{170}{420}$ **b** $\frac{2}{420}$ **c** $\frac{1}{420}$

7 A suitable response section would be: numerical scale 1 to 5, or Very good, Good, Average, Poor, Very poor etc.

8 a and **b**

	1	2	3	4
Red		|		|
Black		||		
White	|			

9 24 times.

10 a 40 houses **b** 4 bedrooms **c** Median = 4 bedrooms. Mean = 3.7 bedrooms (to 1 d.p.).

11 a and **c**

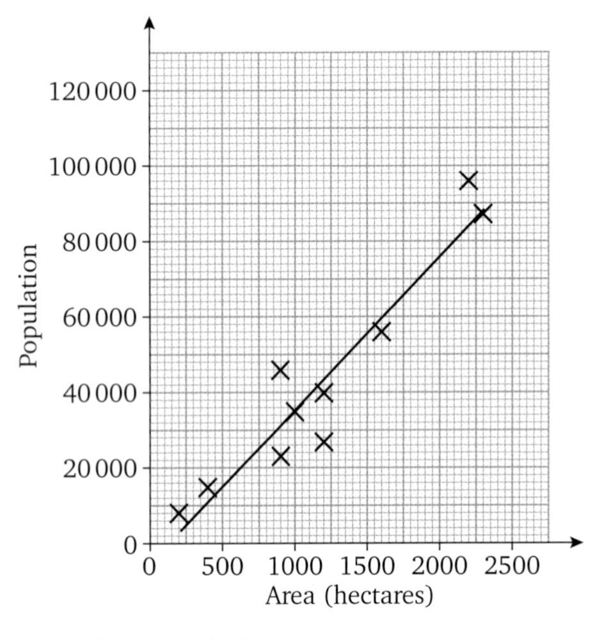

b Quite strong positive correlation.

d Answer between 750 and 850 hectares.

12 a

0	8
1	2 5 6 7 9
2	2 5 5 6 7 8 8
3	2 3 5 6 9
4	0 1

Key: 1 | 2 represents 12 lengths

b Median before = 26.5 lengths. Median after = 26 lengths. Therefore median has decreased.

13 On average boys jump further.

Boys' jumps have a slightly larger range of distances.

Furthest boy jumps further than best girl.

14 No. Girls mean is £5.70 and Boys mean is £6.93.

15 Student's own answer. Must include a description of how the data will be collected, an example of at least one calculation/diagram, and how they will use this to make a conclusion.

Algebra

3 Answers

1 Sequences and symbols

1 a 26, 32 **b** +6

2 a 20 **b** 0 **c** 14

3 a $4a - 3b$ **b** $y + 5z$ **c** $6pq - 2p$

4 4, 9, 14

5 a $8k - 4$ **b** $6m - m^2$ **c** $6t^2 + 4t$

6 a $5(2x + y)$ **b** $3k(k - 3)$

7 a $3n - 1$ **b** $2\frac{1}{2}n - 1\frac{1}{2}$ or $\frac{1}{2}(5n - 3)$

8 $a + 14b$

9 43, 124

10 $a = 4$

11 a True **b** False **c** True **d** False

12 $a = 7$ $b = 2$

2 Equations, inequalities and formulae

1 a £8.50 **b** 8 miles

2 a $m = x + 5$ **b** $n = 2x$

3 a $a = 6$ **b** $b = -1$ **c** $c = 4$ **d** $d = 4\frac{1}{2}$

4 a $p = 4y$ **b** $P = 4m + 4n$

5 $T = 8n + 195$

6 a $e = 18$ **b** $f = 7$ **c** $g = -\frac{1}{2}$ **d** $h = -3$

7 $5x - 7 = 33$ The number is 8.

8 a $k = 5$ **b** $m = -2$ **c** $n = \frac{1}{2}$

9 $3x + 5x + x + 70 + x - 20 = 360$, $x = 31$

10 £$(1\frac{1}{2}x + 2y)$ or $(150x + 200y)$ pence

11 a

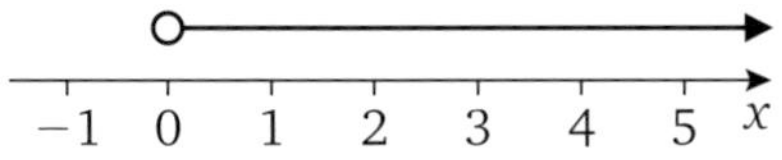

−1 0 1 2 3 4 5 x

b

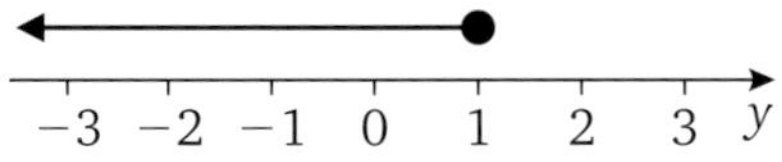

−3 −2 −1 0 1 2 3 y

c

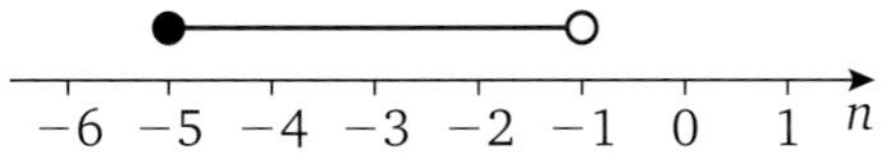

−6 −5 −4 −3 −2 −1 0 1 n

12 a 540° **b** $S = 90(2k - 4)$ **c** $k = 8$

13 a $x \geqslant -1$ **b** $y > -2$

14 −2, −1, 0, 1, 2, 3

15 **a** $x = -6$ **b** $y = 1$ **c** $t = -10$

16 $p = \dfrac{t+7}{3}$ or $\frac{1}{3}(t + 7)$

17 **a** $w = \frac{1}{2}(P - 2l)$ or $\dfrac{P - 2l}{2}$ **b** $6\frac{1}{2}$

18 $3(2x + 15) = 69$ Ryan's number is 4.

19 **a** $x = -6$ **b** $y = 6\frac{1}{2}$ **c** $z = 48$

3 Trial and improvement

1 $x = 1.8$

2 $y = 2.4$

3 $t = 4.1$

4 **a** $m = -2.2$ **b** $m = 2.9$

5 $x = -4.4$

6 $p = 6.8$

7 $y = 3.7$

4 Coordinates and graphs

1 $A(4, 2)$ $B(2, 0)$ $C(3, -3)$ $D(0, -4)$ $E(-2, -5)$ $F(-2, 5)$

2 **a** £40 **b** £115 **c** 2 h 45 min

3 (6, 3) (−6, 3) (2, −5)

4 $y = 5$

x	0	3	6
y	8	4	0

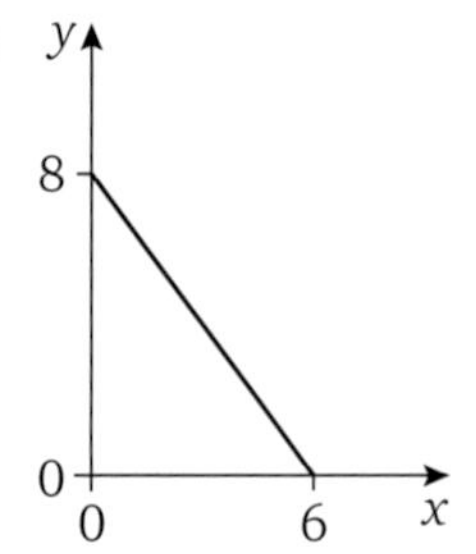

6 **a** 5.2 km **b** He stops for 4 minutes. **c** 6 km/h

d

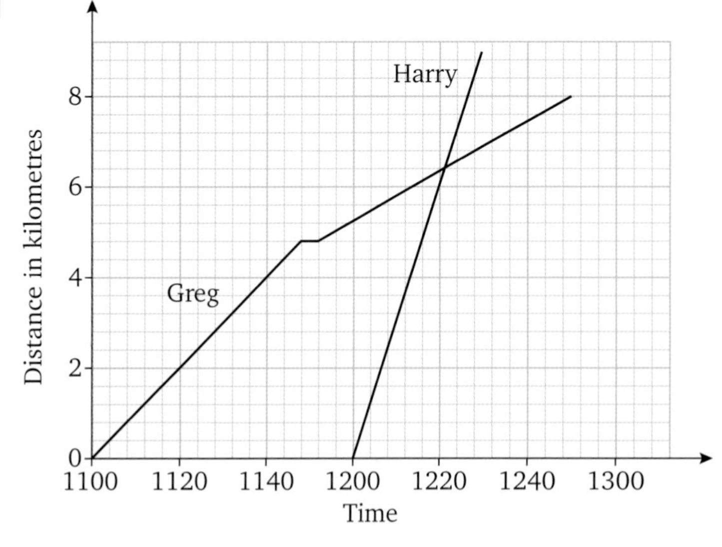

e At 1222, 6.4 km from home.

7 **a** $(1, 2\frac{1}{2})$ **b** (4, 0)

8 $\frac{8}{6}$ or $\frac{4}{3}$

5 Quadratic functions

1 a

x	−3	−2	−1	0	1	2	3
y	14	**9**	6	5	**6**	9	**14**

b

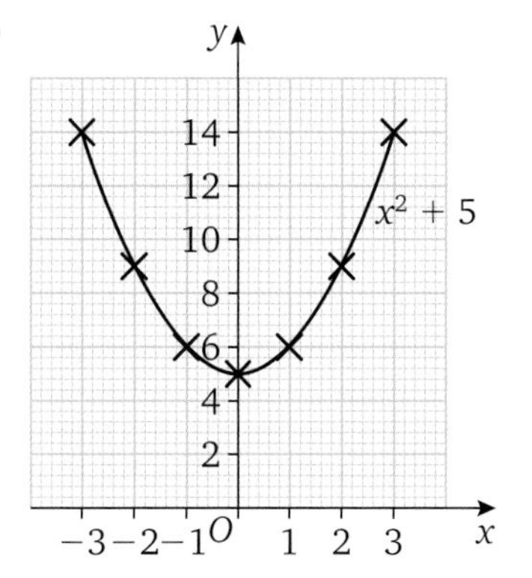

c $x = \pm 1.7$

2 a

x	−2	−1	0	1	2	3	4
y	−10	**−4**	0	2	**2**	**0**	−4

b

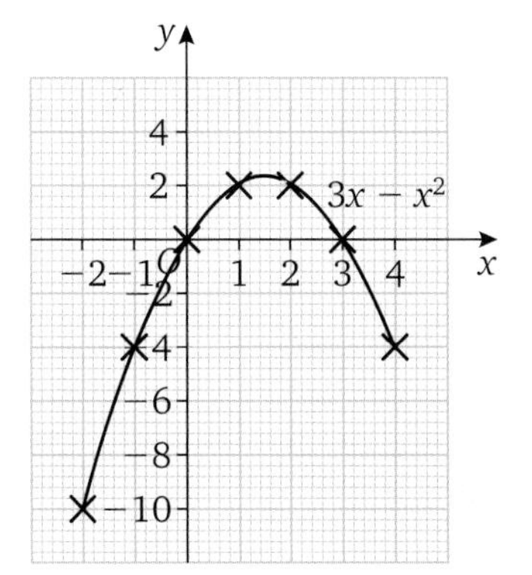

c (1.5, 2.2)

3 a

x	−3	−2	−1	0	1	2	3
y	20	5	**−4**	**−7**	−4	5	**20**

b

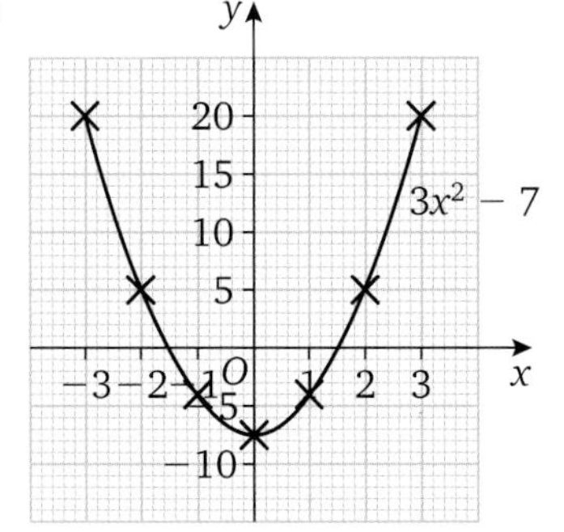

c (1.5, 0) and (−1.5, 0)

4 a, b

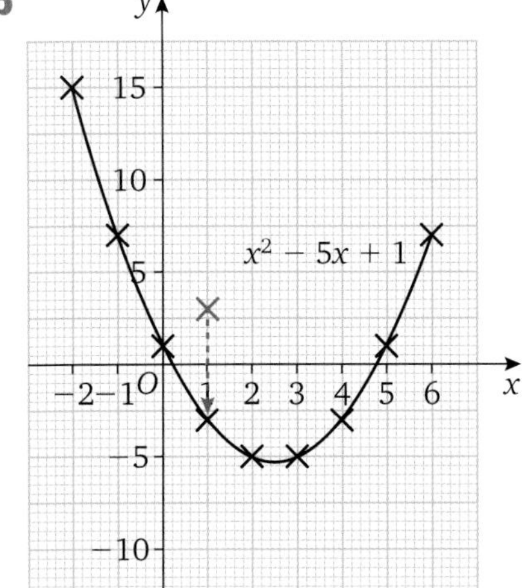

c (2.5, −5.25)

5 a

x	−2	−1	0	1	2	3	4	5
y	7	**0**	−5	−8	**−9**	−8	−5	**0**

b

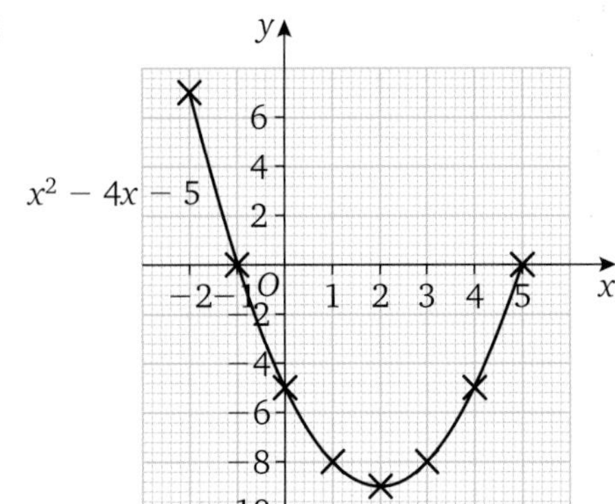

c $x = -0.5$ and 4.5

6 a

x	−1	0	1	2	3	4
y	**−2**	1	2	**1**	−2	−7

b

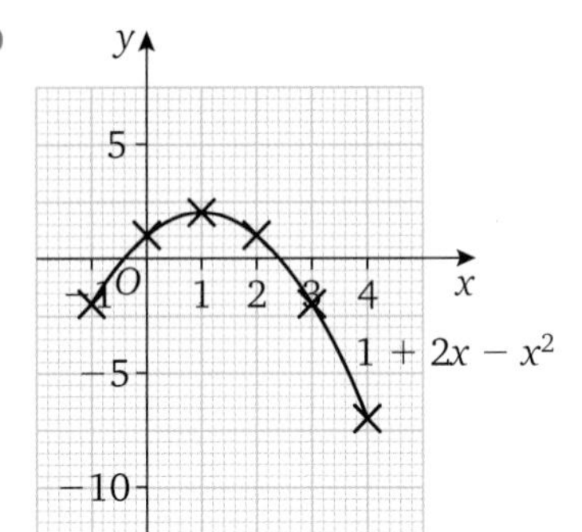

c $x = 1$

6 AQA Examination-style questions

1 a 25, 29 **b** add 4 **c** 41

2 a 5, 2 **b** subtract 3 **c** After this the numbers become negative: −1, −4, etc.

3 a 54 **b** 27

4 a £83 **b** 2 h 45 min

5 a

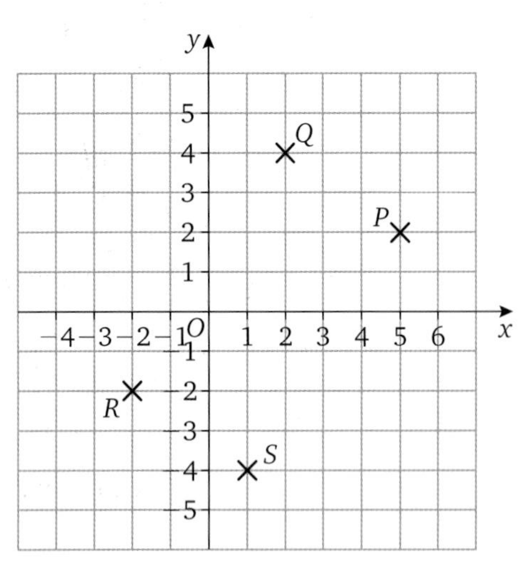

b (1, −4)

6 a $3m$ **b** $5x + 4y$ **c** $3k^2 - 6k + 7$

7 a $x = 3$ **b** $y = -4$ **c** $z = 18$ **d** $t = 3\frac{1}{2}$

8 a 18 m **b** 150 feet **c** 600 feet

9 $6x + 4y$

10 a

x	0	3	5
y	**10**	4	**0**

b and **c**

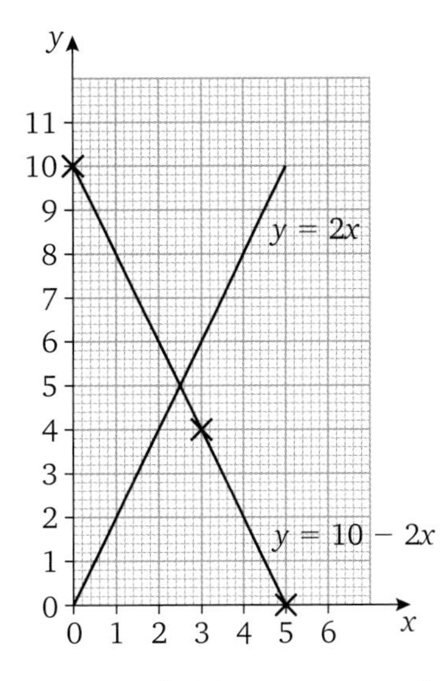

d (2.5, 5)

11 a

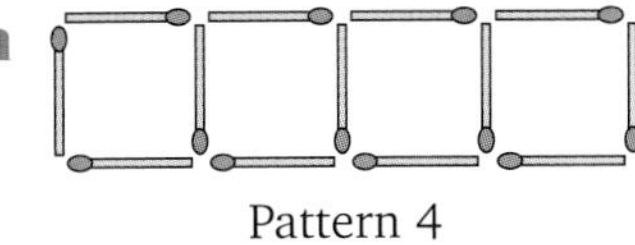

Pattern 4

b 13 **c** 76 **d** 12 **e** $3n + 1$

12 a Second is $x + 5$, third is $2x$ **b** $x + x + 5 + 2x = 61$ **c** $x = 14$

13 a 5 km

b

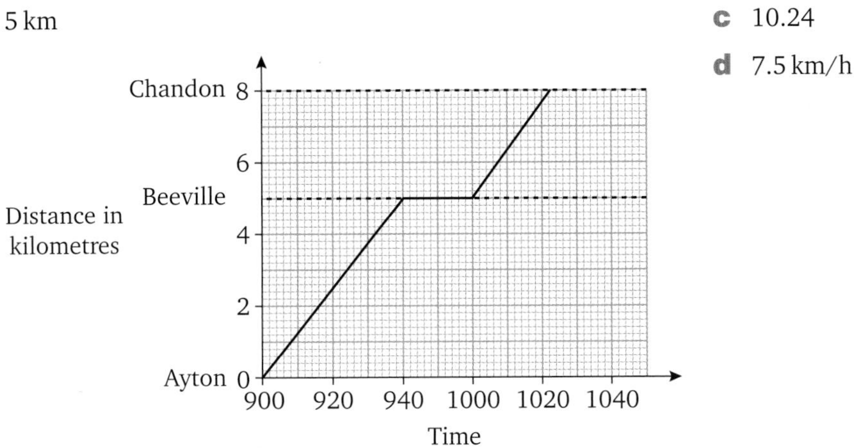

c 10.24

d 7.5 km/h

14 $4x + 25 + 3x + 90 = 360, \ x = 35$

15 a

x	−3	−2	−1	0	1	2	3
y	4	**−1**	−4	−5	**−4**	−1	**4**

b

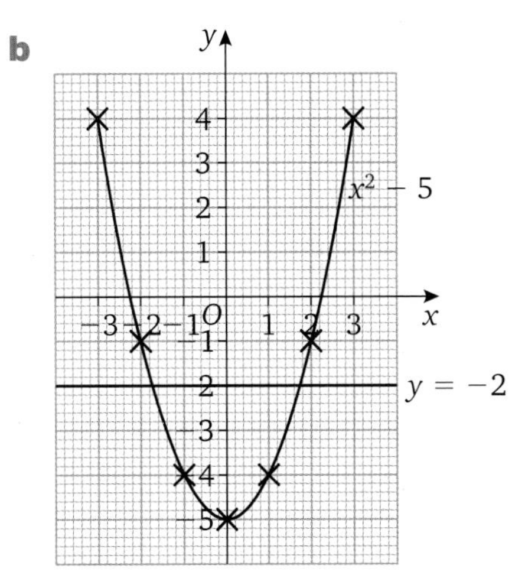

c $x = -1.8$ and 1.8

16 a formula **b** expression **c** formula **d** equation **e** equation **f** expression

17 a $7(x - 3)$ **b** $y(2 - y)$

18 a $p = -2$ **b** $q = -1$ **c** $t = 22$

19 $W = \dfrac{T - 15}{30}$

20 a −1, 0, 1, 2, 3 **b** $y > 2.4$

21 $x = 3.2$

4 Answers

1 Angles

1 **a** y **b** z **c** w **d** x

2 $a = 49°$ $b = 41°$ $c = 139°$

3 105°

4 47°. The exterior angle = the sum of the interior opposite angles.

5 **a** 074° **b** 204° **c** 324°

6 **a** 47° (alternate angles on parallel lines)
b 69° (angles on a straight line)
c 64° (alternate angles on parallel lines)

7 30°

8 105°

9 45°

2 Perimeter, area and volume

1 Area = 12 cm², perimeter = 20 cm

2 Approximately 13 cm²

3 Area = 24 cm², perimeter = 20 cm

4 Area = 30.96 cm², perimeter = 23 cm

5 22.01 cm², 30.38 cm², 39.71 cm², 30.19 cm² (to 2 d.p.)

6 **a** Volume = 192 cm³
b Surface area = 208 cm²

7 Volume = 301.6 cm³ (to 1 d.p.)

8 Width = 4 cm

3 Transformations

1

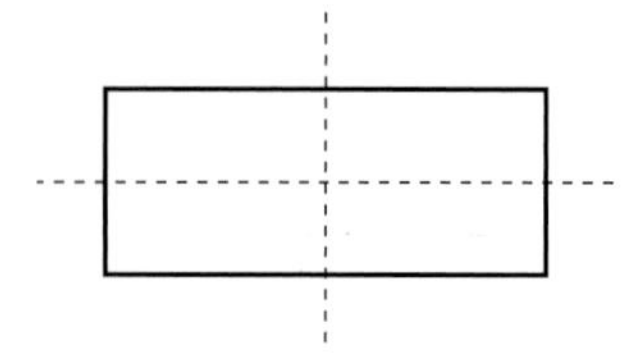

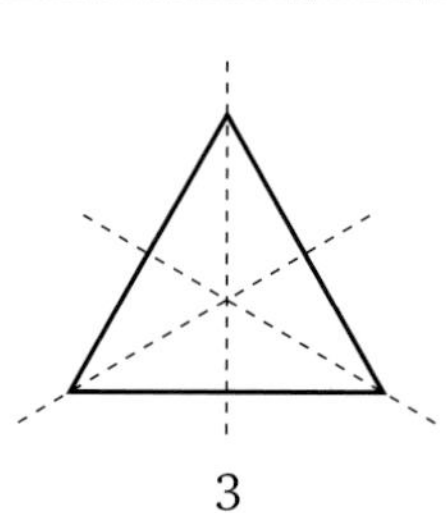

Rotation symmetry order: 2 4 3

2–4

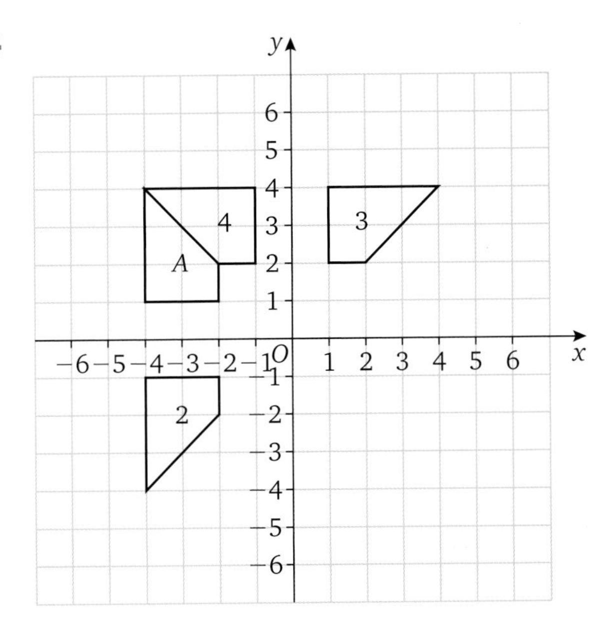

5

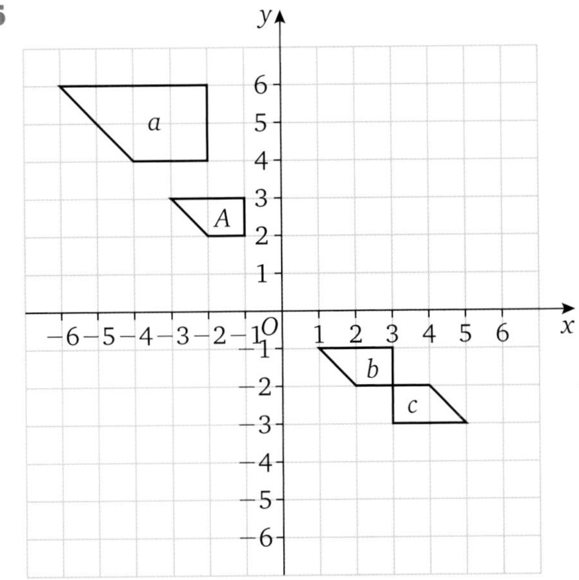

6 a 90° clockwise rotation about (0.5, 3.5)

b Reflection in $y = -1$

c Translation through $\begin{pmatrix} -4 \\ -3 \end{pmatrix}$

d Enlargement, scale factor 2, centre (−5, 4)

e 90° clockwise rotation about (0, 0)

7 a

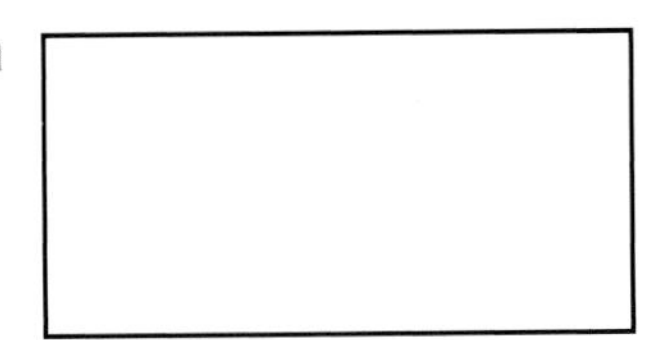

b The rectangle should measure 10 cm × 5 cm

4 Measures, loci, construction

1 a Triangular prism

b *E* and *I*

2 a 1.9 m

b 6 feet

3

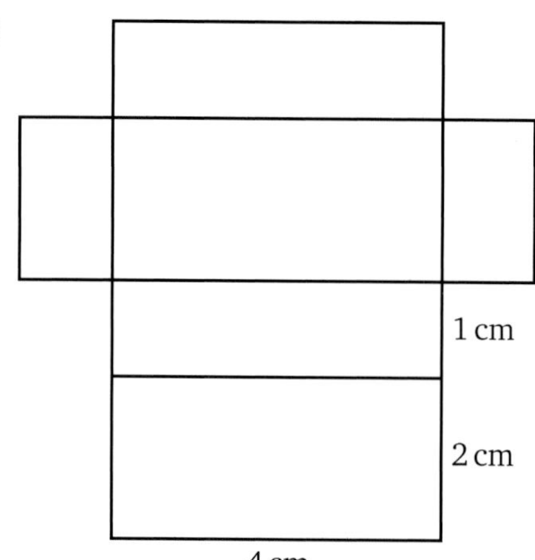

4

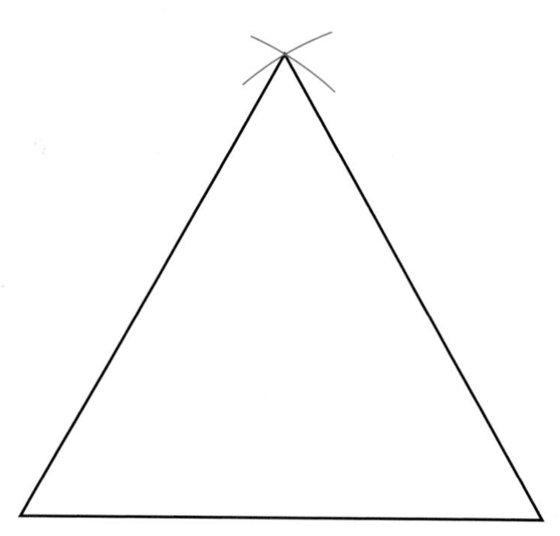

5 7.9 m/s (to 1 d.p.)

6 A

7 a and **b**

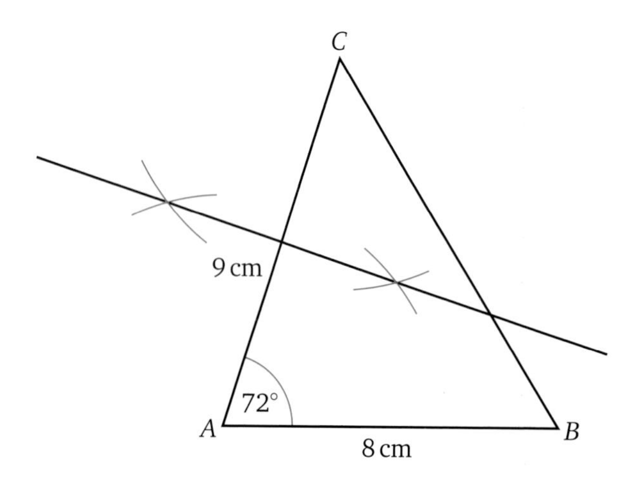

8 a and **b**

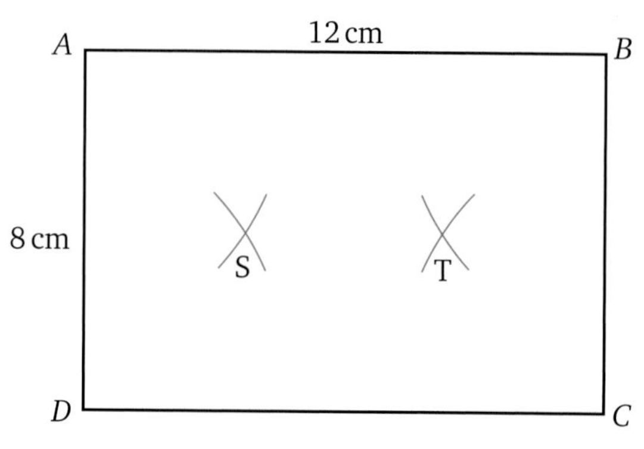

c 30 m

9 a 45 minutes **b** 10 km/h

5 Pythagoras' theorem

1 a 10 cm **b** 6.0 cm (to 1 d.p.) **c** 7.1 cm (to 1 d.p.)

2 6.8 cm (to 1 d.p.)

3 4.1 cm (to 1 d.p.)

4 $AB = 5$ units

5 $5^2 + 12^2 = 169 = 13^2$

6 5.66 m (to the nearest cm)

6 AQA Examination-style questions

1 **a** Sometimes true **b** Never true **c** Always true

2 A, C and D

3 Perimeter = 24 cm Length of square = 6 cm Area = 36 cm^2

4

	1	2	3	4
	5	6	7	
	8	9		

Possible combinations are:

4, 8 and one of 2, 6 or 9 or 3, 4 and one of 5, 6 or 7.

5

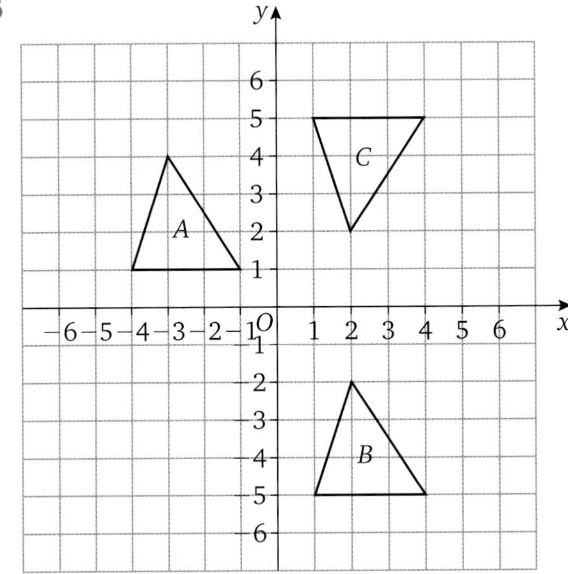

6 $\frac{8 \times h}{2} = 24$

Height = 6 cm

7

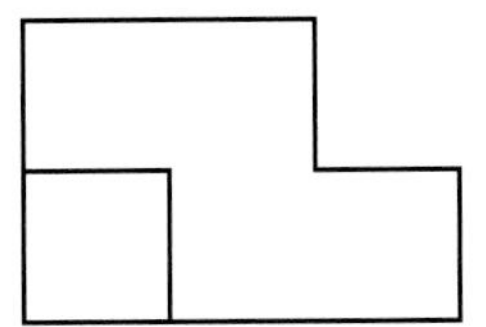

8

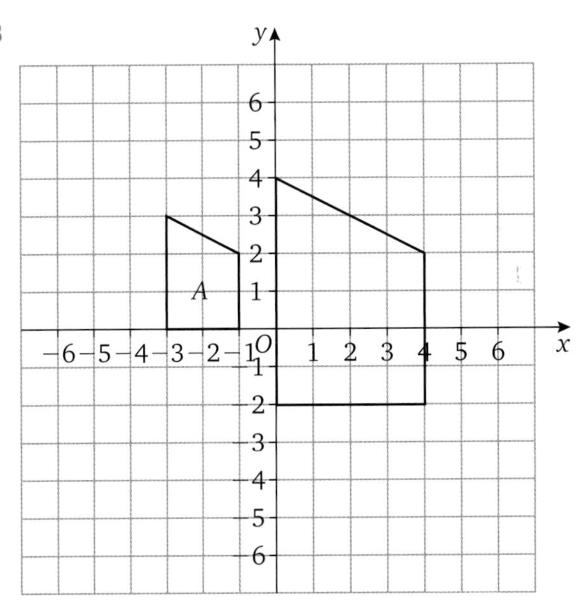

9 a Accurate construction; all lengths to be within 2 mm.

Construction lines as shown:

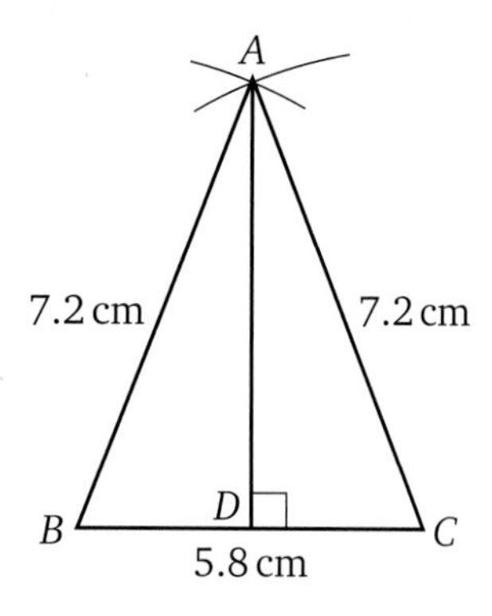

b $BD = 5.8 \div 2 = 2.9\text{ cm}$

$BD^2 + AD^2 = AB^2$

$2.9^2 + AD^2 = 7.2^2$

$AD^2 = 43.43$

$AD = 6.6\text{ cm}$ (to 1 d.p.)

10 Area of triangle $= \frac{1}{2} \times 1.6 \times 1.4 = 1.12\text{ m}^2$

Volume of tent $= 1.12 \times 2.4 = 2.69\text{ m}^3$ (to 2 d.p.)

11 Area of top $= \pi r^2$

$= 314.2\text{ cm}^2$

Area of side $=$ circumference $\times$ height

$= \pi d \times h$

$= 377.0\text{ cm}^2$

Total area $= 691.2\text{ cm}^2$

5 Answers

1 Justifying answers

1 $(42 - 12) \div 5 = 6$ and $18 \div 9 + 4 = 6$ or $(6 - 4) \times 9 = 18$

or

$18 \div 9 + 4 = 6$ and $6 \times 5 + 12 = 32$

2 Rectangle A has a perimeter $5 + 1 + 5 + 1 = 12$ cm

Rectangle B has a perimeter $2 + 3 + 2 + 3 = 12$ cm

Sam is wrong because the rectangles have equal perimeters.

3 The shape cannot be drawn accurately because it is a quadrilateral.

The angles of a quadrilateral add up to 360° and the angles in this shape add up to 357°.

4 Cost: $200 \times 0.8 = £160$

Sales: $0.6 \times 200 \times 3 + 0.4 \times 200 \times 2 = £520$

Profit: $520 - 160 = £360$

Yes, Zoe does manage to make a profit of more than £350

5 Pyramid A: $(n + 2) + (2 + 3n + 1) = 4n + 5$

Pyramid B: $(2n + 1 + 1) + (1 + 2n + 2) = 4n + 5$

6 a Probability of a red $= \frac{1}{6}$

$\frac{1}{6} \times 600 = 100$ compared with 98

Probability of a white $= \frac{1}{3}$

$\frac{1}{3} \times 600 = 200$ compared with 196

Probability of blue $= \frac{1}{2}$

$\frac{1}{2} \times 600 = 300$ compared with 306

Actual outcomes close to theoretical so Chandi is correct – the dice is fair.

b The dice has not been thrown enough times to make any decision about whether it is fair or not.

7 Week 1 – Mean score: 13.36 ... Range: 9

Week 2 – Mean score: 14.6 Range 11

Week 3 – Mean score: 15.66 ... Range 8

On average, scores in week 3 are the highest and most consistent.

On average, scores in week 1 are the lowest.

Scores in week 2 are the least consistent.

8 Outputs: $2x + 8$ and $2(x + 4)$

$2(x + 4) = 2 \times x + 2 \times 4 = 2x + 8$

9 a $2X$ is $2 \times$ an odd number and therefore a multiple of 2 and even.

So $2X + Y$ is an even number plus an even number, which is always even.

b $X + 2Y - 3$ is even because odd (X) + even $(2Y)$ is odd and an odd number – 3 is even.

2 is the only even prime number.

When $X = 1$ and $Y = 2$, $X + 2Y - 3 = 1 + 2 \times 2 - 3 = 2$. So Tim is wrong.

10 a $4(2x + 3) - 6(x + 1) \rightarrow 8x + 12 - 6x - 6 \rightarrow 2x + 6 \rightarrow 2 \times x + 2 \times 3 \rightarrow 2(x + 3)$

b When x is a whole number $(x + 3)$ is a whole number.

So $2 \times (x + 3)$ is $2 \times$ a whole number which is even.

11 Perimeter of square $= 4(2x - 3y) = 8x - 12y$

Perimeter of rectangle $= 2(x + 2y) + 2 \times \text{Width} = 2x + 4y + 2 \times \text{Width} = 8x - 12y$

So $2 \times \text{Width} = 8x - 12y - 2x - 4y = 6x - 16y = 2(3x - 8y)$

So Width $= 3x - 8y$

12 Angle $BDC = 180 - (25 + 30) = 125$
Angle $BDA = 180 - 125 = 55$
Angle BDA = angle BAD, so triangle ABD is isosceles and $AB = BD$

2 Problem solving

1 Work out the saving in June: $305 - (45 + 50 + 55 + 50 + 55) = 50$
Sam saves £50 three times, £55 twice and £45 once.
So the mode is £50.

2 40 minutes.

3 To take 72p from her pocket Charlotte has to take a 50 pence coin, a 20 pence coin and a 2 pence coin.
For this to be the largest possible amount, the other coin must be a 1 pence coin.
So when Charlotte takes three coins from her pocket she can take.
72p (50 + 20 + 2), 71p (50 + 20 + 1), 53p (50 + 2 + 1), 23p (20 + 2 + 1)

4 $2x + 5 + 3x - 8 = 5x - 4 + 4 - x$
$5x - 3 = 4x$
$x = 3$
Area of A = $(2 \times 3 + 5) \times (3 \times 3 - 8) = 11 \times 1 = 11\,\text{cm}^2$
Area of B = $(5 \times 3 - 4) \times (4 - 3) = 11 \times 1 = 11\,\text{cm}^2$
The rectangles have equal area.

5 If x is the number that they both think of $4(x - 3) = 2x + 15$
This gives $x = 13.5$

6 25 minutes.

3 AQA Examination-style questions

1 £1, £1, £1, 50p and 20p or £2, £1, 50p, 10p and 10p.

2 3 A, 4 B and 8 C.

3 Jerry's cuboid has dimensions 15 by 1 by 1 and surface area 62 cm².
Tom's cuboid has dimensions 5 by 3 by 1 and surface area 46 cm².

4 a

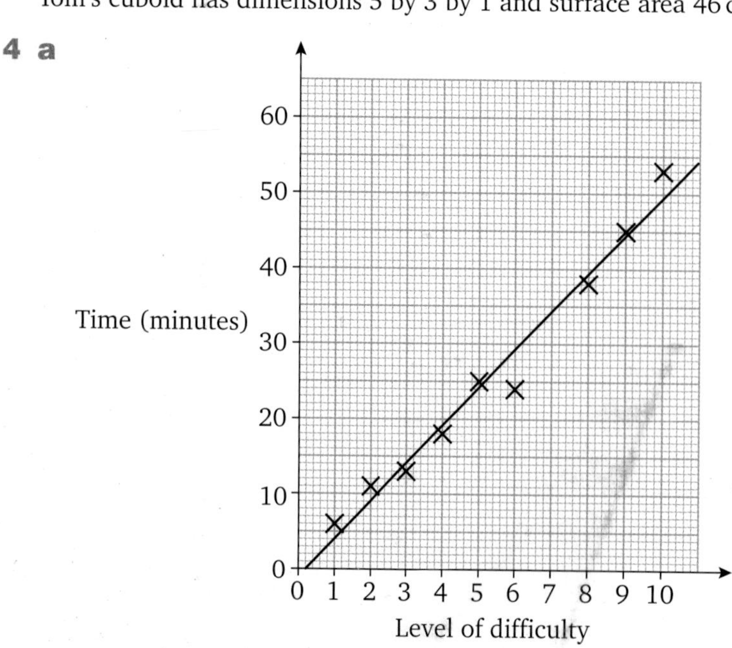

b Use the line of best fit (see graph) to read off the value when $d = 7$. The answer is 34.